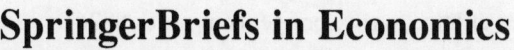

SpringerBriefs in Economics

W0037397

More information about this series at http://www.springer.com/series/8876

Atin Basuchoudhary • James T. Bang • Tinni Sen

Machine-learning Techniques in Economics

New Tools for Predicting Economic Growth

Atin Basuchoudhary
Department of Economics and Business
Virginia Military Institute
Lexington, VA, USA

James T. Bang
Department of Finance, Economics, and
Decision Science
St. Ambrose University
Davenport, IA, USA

Tinni Sen
Department of Economics and Business
Virginia Military Institute
Lexington, VA, USA

ISSN 2191-5504   ISSN 2191-5512   (electronic)
SpringerBriefs in Economics
ISBN 978-3-319-69013-1   ISBN 978-3-319-69014-8   (eBook)
https://doi.org/10.1007/978-3-319-69014-8

Library of Congress Control Number: 2017955621

Printed on acid-free paper

This Springer imprint is published by Springer Nature
The registered company is Springer International Publishing AG
The registered company address is: Gewerbestrasse 11, 6330 Cham, Switzerland

Contents

Chapter 1
Why This Book?

In this book, we develop a Machine Learning framework to predict economic growth and the likelihood of recessions. In such a framework, different algorithms are trained to identify an internally validated set of correlates of a particular target within a training sample. These algorithms are then validated in a test sample.

Why does this matter for predicting growth and business cycles, or for predicting other economic phenomena? In the rest of this chapter, we discuss how Machine Learning methodologies are useful to economics in general, and to predicting growth and recessions in particular. In fact, the social sciences are increasingly using these techniques for precisely the reasons we outline. While Machine Learning itself is not a new idea, advances in computing technology combined with a recognition of its applicability to economic questions make it a new tool for economists (Varian 2014). Machine Learning techniques present easily interpretable results particularly helpful to policy makers in ways not possible with the standard sophisticated econometric techniques. Moreover, these methodologies come with powerful validation criteria that give both researchers and policy makers a nuanced sense of confidence in understanding economic phenomenon.

As far as we know, such an undertaking has not been attempted as comprehensively as here. Thus, we present a new path for future researchers interested in using these techniques. Our findings should be interesting to readers who simply want to know the power and limitations of the Machine Learning framework. They should also be useful in that our techniques highlight what we do know about growth and recessions, what we need to know, and how much of this knowledge is dependable.

Our starting point is Xavier Sala-i-Martin's (1997) paper wherein he summarizes an extensive literature on economic growth by choosing theoretically and empirically ordained covariates of economic growth. He identifies a robust correlation between economic growth and certain variables, and divides these "universal" correlates into nine categories. These categories are as follows:

© The Author(s) 2017
A. Basuchoudhary et al., *Machine-learning Techniques in Economics*,
SpringerBriefs in Economics, https://doi.org/10.1007/978-3-319-69014-8_1

1. Geography. For example, absolute latitude (distance from the equator) is negatively correlated with growth, and certain regions, such as sub-Saharan Africa and Latin America underperform, on average.
2. Political institutions. Measures of institutional quality like strong Rule of Law, Political Rights, and Civil Liberties improve growth, while instability measures like Number of Revolutions and Military Coups and War impede growth.
3. Religion. Predominantly Confucianist/Buddhist and Muslim countries grow faster, while Protestant and Catholic grow more slowly.
4. Market distortions and market performance. For example, Real Exchange Rate Distortions and Standard Deviation of the Black Market Premium correlate negatively with growth.
5. Investment and its composition. Equipment Investment and Non-Equipment Investment are both positively correlated with growth.
6. Dependence on primary products. Fraction of Primary Products in Total Exports are negatively correlated with growth, while the Fraction of Gross Domestic Product in Mining is positively correlated with growth.
7. Trade. A country's Openness to Trade increases growth.
8. Market orientation. A country's Degree of Capitalism increases growth.
9. Colonial History. Former Spanish Colonies grow more slowly.

Sala-i-Martin's findings are standard in the growth literature. His econometric techniques cull the immense proliferation of explanatory variables into a tractable and parsimonious list. However, there are several problems with his approach that in turn hint at fundamental gaps in our understanding of the economic growth process. The Machine Learning framework can fill precisely these kinds of gaps in evidence.

The findings of the standard econometric techniques deployed by Sala-i-Martin cannot say anything about why certain variables matter, or which matter more than others. For example, if a country's GDP has a large *Fraction of Primary Products in Total Exports*, it is likely to be a growth laggard, though if it has a high *Fraction of GDP in Mining*, it is in the high growth category. This sort of contradiction suggests that maybe the Sala-i-Martin list is not parsimonious enough. It is certainly not always amenable to consistent theoretical explanations.

In our treatment, we start with a set of variables and dataset that largely mirrors Sala-i-Martin's comprehensive list of (what he identifies as) robust correlates of economic growth. Next, we randomly pick a set of countries to divide the data set into a learning sample (70% of the data) and a test sample (30% of the data). We use multiple Machine Learning algorithms to find the algorithm with the best out-of-sample fit. We then identify the variables that contribute the most to this out-of-sample fit. Thus, the algorithms can rank variables according to their relative ability to predict the target variable. We can thus whittle down the correlates of growth identified by Sala-i-Martin to the ones that robustly contribute to prediction. Thus, we are able to identify those variables that best predict growth and recessions 5 years out, without any of the inherent contradictions outlined above.

In our analysis, a country in a particular year is the observational unit. We structure the data so that the target (growth or recession) is 5 years out. For example, the first period contains covariates for 1971–1975, while the target is growth, or an incidence of recession, in the 1976–1980 period. Looking at growth in 5-year periods is standard in the literature. However, choosing the dependent variable or target 5-years out is, to our knowledge, new in the literature. This data structure is therefore our first innovation toward developing a truly predictive model. Our targets are economic growth and recessions.

We also report the marginal effect of these variables on economic growth and recessions through partial dependence plots or PDPs. The PDPs provide insights on the pathways of economic growth. They tell us how changing a variable affects the target over the range of that change. Thus, we are able to say (with some sense of the confidence that comes from estimates of predictive accuracy) whether, over a certain range, a particular variable has a greater or lesser effect on growth, whether it affects growth negatively or positively, as well as identify other ranges where the variable does not affect growth. Thus, if we find that Investment is an important predictor of growth, the PDP shows us how an increase in investment affects growth over the range of that increase. In fact, we find that the covariates of growth affect growth in consistently non-linear ways. A parametric point estimate cannot capture this non-linearity. The information in PDPs is particularly useful to policy makers, when, for instance, it comes to understanding how countries with different levels of investment may respond differently to changes in a policy lever. It also has implications for the process of developing theoretical models of growth in that these models need to take into account these non-linearities.

The growth literature's focus on growth accounting and regressions, and therefore on the correlates of growth, ends up generating long lists of possible correlates of growth. Such lists hamper standard econometric techniques since they are plagued by a number of problems—parameter heterogeneity, model uncertainty, the existence of outliers, endogeneity, measurement error, and error correlation (Temple 1999), to name a few. In the following chapters, we suggest that Machine Learning can help circumvent some of these problems. Thus, Machine Learning methodologies that create parsimonious lists of the covariates of growth that are validated by out-of-sample fit can be particularly useful in the growth literature. They can complement current econometric methodologies, and, at the same time, they can offer fresh insights into economic growth.

Standard econometric techniques, the only ways to discern causality in pathways to growth and away from recessions, require assumptions about underlying distributions for them to even be valid within a sample, let alone ever be tested out-of-sample. Further, the variables that are used in these statistical models arise out of (mathematically) internally consistent models. However, there is no clear way to know which of these may actually be a theory of growth. For example, is the Solow approach to growth a better contender for a theory of growth than Romer's endogenous growth models? This of course begs the question, what influences these theoretical models—technology, institutions, culture, and so on. The list is endless since model specifications along these lines are only limited by the infinite human capability of thought. Machine Learning has the advantage of not requiring any prior assumptions about theoretical links, or indeed any major assumptions about a variable's underlying distribution.

Why do we bring so much attention on the Machine Learning framework's ability to validate out-of-sample? It is because good theory should identify causal pathways to explain phenomenon, and such causal pathways should be generalizable. Further, the test of such generalizability is in the theory's ability to *predict* a relevant phenomenon. So, a theory of gravity that explains reality only in New York and cannot predict the effects of gravity elsewhere, is not really a theory of gravity. Thus, a theory of growth that cannot predict growth is not really a theory of growth. Machine Learning algorithms are by definition validated out of sample, i.e. they are predictive. These algorithms are therefore uniquely poised to check whether growth theories are generalizable by scoring predictive accuracy.

Variables that appear to be robust correlates of growth but do not predict well out-of-sample cannot really be causal variables since they are not generalizable outside of a particular sample. In such cases, the patterns among these variables are mere anecdotes. Thus, eliminating variables that do not contribute to a model's out-of-sample fit help us focus only on variables that can be shown to maximize out-of-sample fit, i.e. they are generalizable. We suggest that the search for causal theories of growth should begin among the pathways of influence suggested by these variables. Machine Learning can therefore be helpful in exploring causal links to growth (Athey and Imbens 2015).

The process of variable elimination can also help distinguish between different theories of growth. Indeed, Machine Learning algorithms appear to identify a particular theoretical strand (model) as most salient based on its out-of-sample fit. To the extent that our target is growth or recessions 5 years out, the extent of this salience also informs us about the extent of the generalizability of this theoretical strand. By leveraging Machine Learning to score different theoretical paradigms on predictive quality, we offer a consistent methodology for judging how much confidence we should place on theoretical models. We suggest this approach should become standard in the absence of randomized controlled trials.

Machine Learning algorithms are atheoretical. However, a researcher can choose variables to include in an algorithm. The algorithms constantly sample and resample within the training sample to come up with models that fit the data best. These models are then validated out-of-sample. Apart from the initial choice of variables, the entire process is untouched by human hands. Nevertheless, the hands-free process tells us whether that initial choice of variables was valid or not in a very simple way—through out-of-sample fit. To the extent, the test sample is chosen randomly this process also helps reduce researcher bias. We recommend it for that reason as well.

Machine Learning techniques have practical benefits as well. For instance, the policy maker mainly needs to know the effect of a current change in policy on a future (out-of-sample) target. From the policy maker's perspective, a list of variables identified by the usual econometric techniques do not provide good policy levers for increasing economic growth because these studies tend to neglect out-of-sample predictability. For example, the policy maker has no idea whether s/he should focus on reducing inflation or on spending more on healthcare to induce higher growth.

Econometric techniques suggest that both inflation-reduction and increased healthcare expenditure are correlated with growth. They also provide parametric estimates of marginal effects. However, these techniques typically are not validated in terms of out-of-sample predictive ability. Machine Learning, on the other hand, emphasizes out-of-sample prediction scores for different model specifications to predict growth. Additionally, some algorithms rank variables based on how much they individually contribute to out-of-sample fit. This distinction between econometric approaches and Machine Learning approaches matters. For example, say econometrics suggests that inflation has a larger marginal effect on growth than healthcare spending. Machine Learning algorithms on the other hand suggests that healthcare investment contributes more to predicting growth out-of-sample than inflation. From a policy perspective then inflation is less likely to influence future (out-of-sample) growth than healthcare investment. Thus, comparing magnitudes of parametric point estimates to implement policy may be misleading. Policy makers can use Machine Learning to prioritize policy levers according to which ones may have the greatest impact on economic growth.

Moreover, even the robustness of the in-sample correlation is suspect because the techniques themselves are sensitive to assumptions about the underlying distributions of the variables. As a result, current common econometric empirical approaches do not give policy makers a sense of how much reliance they can place on these results.

Another problem in the growth literature is the paucity and unreliability of data for precisely the countries for which growth issues matter most. Standard statistical analyses do not perform well when there is missing data. Machine Learning can address this problem in a scientifically verifiable way by finding "surrogate" variables that can proxy those with missing data. These proxies are chosen by the Machine Learning techniques by their predictive abilities, and to that extent, provide a hard test for the usefulness of a particular proxy variable.

We plan to develop a framework for understanding the complex non-linear patterns that link formal political institutions, informal political institutions, resource availability, and individual behavior to economic growth. Our empirical strategy atheoretically incorporates the patterns that link underlying variables to predict the rate of economic growth. We repeat this to predict the likelihood of recessions. In both cases, we provide the reader with the criterion for judging the validity of our results. In the process, we note gaps in our current understanding of growth and suggest future directions of research. Then, we take those factors that our empirical model identifies as important, and suggest a roadmap to build a theoretical framework that explains how these fit into the story of growth. For both growth and recessions, we identify those variables with the most salience for policy makers that are rooted in the current literature. This literature may have gaps but policy cannot wait for settled science. Policy makers need to make the best possible decisions with the information they have. We posit a framework to identify the "best" among the policy levers we know of. We cannot do anything about the unknown unknowns. Thus, we have two goals in this book. One goal is to show how Machine Learning can help highlight evidence gaps that econometric techniques cannot. Our pathway to this first goal suggests that our current understanding of economic growth has significant evidence gaps. This finding implies that despite the centrality of economic growth to the economics profession, much of our understanding may be incomplete. Nevertheless, by highlighting the evidence gaps we shed light on how to advance our knowledge of the drivers of economic growth. The second is to highlight how policy makers can use Machine Learning to develop criteria to make better, more nuanced, policy decisions. Our pathway to this second goal suggests that policy makers need to be humble about the effectiveness of any growth policy.

We describe our data in Chap. 2. Chapter 3 describes the algorithms we use. In Chap. 4, we discuss criteria for choosing algorithms and how these choices resolve some endemic problems in the growth literature. In Chap. 5, we show how we can use Machine Learning to sift through different pathways of economic growth to identify the one that matters the most. We discuss what this kind of identification means for causal inferences while noting that prediction and causality are not the same thing. We reevaluate the framework we advocate in Chaps. 4 and 5 by attempting to predict recessions in Chap. 6. We collate the main takeaways from each chapter in our epilogue. The reader interested in future research will find a comprehensive documentation of R codes we have used for this book in the Appendix. Some of the data we use are proprietary and therefore cannot be released publicly. However, we are happy to provide the dataset for replication purposes only. Any further research using this dataset requires the researcher to buy some components from the sources we cite.

Our narrow focus here is to show how Machine Learning can help develop a framework that allows a better understanding of growth and business cycles. Thus, we try to sketch the broad sweep of the literature rather than positing a comprehensive state of the art review of the growth or business cycle literature. Nevertheless, we hope that this book will be useful both to those who want to advance their research using the techniques we apply here as well those who just want a birds-eye view of both the power and limitations of the current understanding of growth through a Machine Learning lens. We suggest the former read the entire book including the R Appendix. The latter can get by with reading Chaps. 2 and 4–6. Readers interested only in growth may want to read Chaps. 2, 4, and 5 and those interested in only recessions can get by with reading Chaps. 2 and 6. We provide the intuition behind the methodologies we use in each chapter. Therefore, these chapters can really be "stand-alone" reads, with reference to the Table of variables in Chap. 2.

References

Athey, S., & Imbens, G. (2015). *Machine learning methods for estimating heterogeneous causal effects.* arXiv Preprints, 1–9. Retrieved from http://arxiv.org/pdf/1504.01132v2.pdf

Sala-i-Martin, X. (1997). I just ran four million regressions. *American Economic Review, 87,* 178–183.

Temple, J. (1999, March). The new growth evidence. *Journal of Economic Literature, 37,* 112–156.

Varian, H. (2014). Big data: New tricks for econometrics. *Journal of Economic Perspectives, 28* (2), 3–28.

Chapter 2
Data, Variables, and Their Sources

In this chapter, we describe our data, explain the need for 'preparing' the data, and finally, describe the process by which we prepare the data. Briefly, our data are from the 2014 Cross-National Time Series (CNTS) 2012 Database of Political Institutions (DPI), International Country Risk Guide (ICRG), Political Instability (formerly, State Failure) Task Force (PITF) and the World Development Indicators (WDI), over the period 1971–1974.

Several of our variables may have similar sources of variation. This is particularly true of institutional variables and those purporting to capture social and political aspects of a country. For example, democratic countries may also have liberal and inclusive economic institutions as well as a better sense of the rule of law than autocratic countries. All of that may have an effect on ethnic conflict. This overlap between different sociopolitical measures makes interpreting them as separate entities difficult. We use EFA techniques to identify unique dimensions among such variables.

Country level data is also problematic because some data may be missing. Indeed, this missing data problem may be particularly problematic for precisely those poor countries where policy can have the biggest impact. We use a validated imputation technique to help mitigate that problem.

We describe our data in Sect. 2.1. We then introduce the EFA technique in Sect. 2.2 and the imputation technique in Sect. 2.3. Summary statistics for the learning and test samples of our raw and imputed datasets are in Tables 2.1 and 2.2. We summarize our variables by category in Table 2.3 and report the EFA results in Table 2.4.

© The Author(s) 2017
A. Basuchoudhary et al., *Machine-learning Techniques in Economics*,
SpringerBriefs in Economics, https://doi.org/10.1007/978-3-319-69014-8_2

Table 2.1 Summary statistics for raw data

Source	Learning sample					Test sample				
	Obs.	Mean	Std. dev.	Min	Max	Obs.	Mean	Std. dev.	Min	Max
GDP pc growth	5388	1.895	4.342	−42.600	51.920	2322	2.177	4.480	−33.500	58.200
Lag GDP pc	4762	831.697	2099.474	−24940.000	28000.000	1973	832.641	1936.719	−3900.000	22,100.000
Lag GDP pc growth	4723	0.123	5.024	−38.150	57.066	2016	0.136	5.034	−46.686	37.587
Aid & dev. asst.	3875	41,000,000	521,000,000	−4,970,000,000	7,090,000,000	1795	64,100,000	650,000,000	−10,400,000,000	11,800,000,000
Consumption/GDP	4364	−0.553	9.929	−133.600	181.120	1829	−0.105	7.631	−39.925	41.980
Dependency	4838	−2.161	4.358	−31.380	18.160	2046	−2.979	4.606	−21.960	14.380
Export prices	2955	56.979	157.584	−410.200	4820.000	1359	51.624	97.053	−111.260	931.920
Exports/GDP	4608	1.682	7.381	−38.366	58.000	1881	1.390	9.014	−50.460	58.420
FDI/GDP	3692	0.666	5.409	−56.184	97.240	1645	0.838	3.634	−26.576	32.150
Fuel exports/GDP	3493	1.187	9.643	−81.061	87.770	1510	0.517	7.992	−88.100	72.471
Fuel imports/GDP	3600	0.768	6.332	−52.833	23.319	1606	0.603	5.805	−46.268	37.865
Gini coefficient	2690	0.055	3.030	−16.655	15.018	1160	−0.162	3.113	−21.002	15.396
Government/GDP	4492	−0.021	4.024	−72.920	39.933	1795	0.253	3.313	−19.818	24.003
Growth	4087	−20.369	369.406	−12,301.080	5026.260	1832	−10.415	340.905	−4949.040	4877.680
Import prices	2955	55.322	87.900	−164.933	805.600	1359	50.389	76.799	−131.100	518.200
Imports/GDP	4608	1.493	13.998	−180.280	273.653	1881	1.253	9.096	−35.000	46.050
Industry/GDP	3785	0.281	4.735	−39.440	22.360	1594	−0.033	4.854	−17.960	47.160
Inflation	3781	−7.922	339.607	−6218.580	6446.500	1645	−12.503	228.239	−4455.040	2303.300
Interest rate spread	2630	−1.728	25.780	−765.440	280.686	1111	−30.362	522.007	−14,699.060	684.260
Investment	4352	0.430	7.496	−90.480	84.850	1832	0.010	5.749	−29.220	26.490
Lending interest rate	2844	−4.420	46.467	−1280.620	231.500	1148	−246.255	4355.439	−121,872.500	1439.900
Life expectancy	4870	1.530	1.608	−16.140	15.140	2079	1.433	1.471	−8.400	9.120
Military/GDP	2020	−0.305	2.307	−35.688	32.470	882	−0.316	0.905	−5.790	5.409
Mineral rents/GDP	4216	0.113	1.675	−16.086	30.840	1783	0.106	1.917	−18.314	16.579
Money/GDP	4117	4.492	11.935	−129.200	89.400	1838	−13.507	295.508	−6916.880	93.000
Phones/population	4575	1.974	3.762	−18.240	20.440	2025	1.853	3.377	−13.400	17.320

Population growth	4991	−0.051	1.026	−9.088	12.720	2145	−0.075	0.747	−4.230	6.730
Real interest rate	2776	1.052	10.077	−73.017	89.380	1129	0.858	39.305	−846.991	416.320
Rural population	4991	−2.074	2.148	−14.660	9.900	2144	−1.627	1.868	−9.500	2.640
Secondary enrollment	3453	4.440	7.237	−43.400	43.825	1571	5.052	7.984	−44.400	54.200
Terms of trade	2285	0.280	28.202	−164.600	104.200	955	−1.463	33.681	−362.950	88.460
Total population	4991	3,020,207	10,400,000	−3,400,000	105,000,000	2145	1,043,260	2,186,589	−2,320,000	17,200,000
Trade	4608	3.173	18.080	−181.200	285.820	1881	2.645	16.624	−70.000	93.800
Democracy	1514	0.112	0.426	−2.333	2.783	604	0.159	0.470	−1.998	2.448
Transparency	1514	−0.027	0.393	−2.217	2.043	604	−0.043	0.397	−1.831	1.711
Ethnic violence	1514	0.033	0.486	−1.868	2.786	604	0.063	0.472	−1.543	1.824
Protest	1514	0.005	0.575	−5.641	4.543	604	0.027	0.316	−1.063	2.841
Regime	1514	0.025	0.540	−3.061	3.452	604	0.045	0.560	−2.508	2.796
Within	1514	−0.015	0.516	−2.480	2.090	604	−0.014	0.468	−3.503	1.799
Credibility	1514	0.161	0.541	−1.535	2.277	604	0.185	0.543	−1.594	1.730

Table 2.2 Summary statistics for random forest imputed data

Source	Learning sample					Test sample				
	Obs	Mean	Std. dev.	Min	Max	Obs	Mean	Std. dev.	Min	Max
GDP pc growth	5388	1.895	4.342	−42.600	51.920	2322	2.177	4.480	−33.500	58.200
Lag GDP pc	5388	867.370	1977.063	−24,940.000	28,000.000	2322	884.323	1790.693	−3900.000	22,100.000
Lag GDP pc growth	5388	−0.186	4.794	−38.150	57.066	2322	−0.216	4.790	−46.686	37.587
Aid & dev. asst.	5388	34,400,000	442,000,000	−4,970,000,000	7,090,000,000	2322	53,500,000	572,000,000	−10,400,000,000	11,800,000,000
Consumption/GDP	5388	−0.342	8.981	−133.600	181.120	2322	0.004	6.826	−39.925	41.980
Dependency	5388	−2.159	4.131	−31.380	18.160	2322	−2.884	4.333	−21.960	14.380
Export prices	5388	55.047	120.318	−410.200	4820.000	2322	52.791	78.542	−111.260	931.920
Exports/GDP	5388	1.524	6.848	−38.366	58.000	2322	1.296	8.130	−50.460	58.420
FDI/GDP	5388	0.648	4.486	−56.184	97.240	2322	0.768	3.073	−26.576	32.150
Fuel exports/GDP	5388	1.047	7.782	−81.061	87.770	2322	0.607	6.460	−88.100	72.471
Fuel imports/GDP	5388	0.805	5.197	−52.833	23.319	2322	0.701	4.852	−46.268	37.865
Gini coefficient	5388	0.015	2.154	−16.655	15.018	2322	−0.098	2.210	−21.002	15.396
Government/GDP	5388	0.071	3.684	−72.920	39.933	2322	0.299	2.922	−19.818	24.003
Growth	5388	−21.691	322.020	−12,301.080	5026.260	2322	−14.196	303.151	−4949.040	4877.680
Import prices	5388	54.235	69.346	−164.933	805.600	2322	52.244	62.446	−131.100	518.200
Imports/GDP	5388	1.458	12.949	−180.280	273.653	2322	1.276	8.193	−35.000	46.050
Industry/GDP	5388	0.211	3.995	−39.440	22.360	2322	0.008	4.052	−17.960	47.160
Inflation	5388	−7.488	285.602	−6218.580	6446.500	2322	−11.163	193.563	−4455.040	2303.300
Interest rate spread	5388	−6.653	23.504	−765.440	280.686	2322	−19.629	361.443	−14,699.060	684.260
Investment	5388	0.324	6.766	−90.480	84.850	2322	0.017	5.134	−29.220	26.490
Lending interest rate	5388	−43.949	132.578	−1280.620	231.500	2322	−155.796	3065.645	−121,872.500	1439.900
Life expectancy	5388	1.532	1.529	−16.140	15.140	2322	1.446	1.393	−8.400	9.120
Military/GDP	5388	−0.334	1.417	−35.688	32.470	2322	−0.337	0.564	−5.790	5.409

Mineral rents/GDP	5388	0.113	1.485	−16.086	30.840	2322	0.106	1.682	−18.314	16.579
Money/GDP	5388	3.548	11.421	−129.200	89.400	2322	−10.442	262.991	−6916.880	93.000
Phones/population	5388	1.990	3.471	−18.240	20.440	2322	1.902	3.158	−13.400	17.320
Population growth	5388	−0.045	0.988	−9.088	12.720	2322	−0.067	0.719	−4.230	6.730
Real interest rate	5388	1.001	7.325	−73.017	89.380	2322	0.873	27.426	−846.991	416.320
Rural population	5388	−2.060	2.068	−14.660	9.900	2322	−1.646	1.797	−9.500	2.640
Secondary enrollment	5388	4.576	5.822	−43.400	43.825	2322	5.052	6.578	−44.400	54.200
Terms of trade	5388	−0.871	19.119	−164.600	104.200	2322	−1.705	22.125	−362.950	88.460
Total population	5388	2,951,327	10,000,000	−3,400,000	105,000,000	2322	1,121,556	2,119,833	−2,320,000	17,200,000
Trade	5388	2.981	16.739	−181.200	285.820	2322	2.574	14.982	−70.000	93.800
Democracy	5388	0.112	0.239	−2.333	2.783	2322	0.114	0.251	−1.998	2.448
Transparency	5388	−0.020	0.225	−2.217	2.043	2322	−0.011	0.217	−1.831	1.711
Ethnic violence	5388	0.032	0.264	−1.868	2.786	2322	0.040	0.247	−1.543	1.824
Protest	5388	0.008	0.313	−5.641	4.543	2322	0.013	0.174	−1.063	2.841
Regime	5388	0.051	0.332	−3.061	3.452	2322	0.051	0.328	−2.508	2.796
Within	5388	−0.016	0.278	−2.480	2.090	2322	−0.017	0.243	−3.503	1.799
Credibility	5388	0.151	0.336	−1.535	2.277	2322	0.166	0.328	−1.594	1.730

Table 2.3 Variables by category

Persistence and convergence effects	Lagged real per capita GDP growth and lagged real per capita GDP
Composition of domestic output and expenditure	Consumption, investment, industry, total trade, imports, exports, mineral rents, fuel imports, fuel exports, foreign direct investment, government expenditure, military expenditures, and foreign aid/development assistance. All as share of GDP
Technology diffusion	Number of phones, as a percentage of the total population
Domestic monetary and price factors	Money supply (As share Of GDP), the rate of growth in money supply, the cpi inflation rate, the lending interest rate, the real interest rate, and the interest rate spread, the terms of trade, the export price index, and the import price index
Demographics and human development	The total population, the population growth rate, the rural population as a percentage of the total population, the dependency ratio (measured as the ratio youth aged 0–15 and elderly aged 65 and over to the working-age population aged 16–64), life expectancy, and the gross secondary school enrollment rate, gini coefficient measure of income inequality
Institutional measures (from EFA)	Democracy, violence, transparency, protest, within-regime instability, credibility, regime instability

2.1 Variables and Their Sources

The target variable in our analysis is the 5-year moving average of the yearly growth rate of real per capita gross domestic product (GDP). We have taken this variable from the World Development Indicators. We use this variable as our main proxy for increases in economic output and well-being. The data for the all models we run are from 1971–2014. The first period contains predictors for 1971–1975 to predict growth in the 1976–1980 period; the last period contains predictors from 2005–2009 to predict growth in the 2010–2014 period.

In slight contrast to Sala-i-Martin's rather exhaustive search for robustly significant covariates, our goal is to focus on potential "policy levers." Thus, we omit the various fixed effects like geography, religion and colonial origin that some studies have found to matter. After all, a country cannot easily change its location, history, or religion! Instead, we begin with a list of those time-variant inputs that other studies have found to be important in explaining growth. The first two of these variables, lagged real per capita GDP growth and lagged real per capita GDP, proxy for the persistence effects of past growth and convergence effects, respectively. From there, we add several variables relating to the composition of domestic output and expenditures: Consumption, investment, industry, total trade, imports, exports, mineral rents, fuel imports, fuel exports, foreign direct investment, government expenditure, military expenditures, and foreign aid/development assistance. Each of these variables is measured in terms of its share of GDP and is lagged in a way similar to the lagged values of GDP per capita and its growth rate. We add the number of phones as a percentage of the total population as a measure of the level of, and penetration of, technology in the economy.

Next, we include lagged values of several variables to account for domestic monetary and price factors that may affect growth: the money supply as a share of GDP, the rate of growth in the money supply, the CPI inflation rate, the lending interest rate, the real interest rate, and the interest rate spread. We add several additional factors that capture the impacts of external forces on price levels: the terms of trade, the export price index, and the import price index. These variables also come directly from the World Development Indicators.

We also include several lagged variables pertaining to demographics and human development. The variables in the WDI from this category are; the total population, the population growth rate, the rural population as a percentage of the total population, the dependency ratio (measured as the ratio youth aged 0–15 and elderly aged 65 and over to the working-age population aged 16–64), life expectancy, and the gross secondary school enrollment rate. To these, we add the Gini coefficient measure of income inequality, which we have obtained from the Standardized World Income Inequality Database (SWIID) compiled by Solt (2016).

Finally, we consider variables that capture various aspects of institutional quality and stability from the 2014 Cross-National Time Series (CNTS) 2012 Database of Political Institutions (DPI), International Country Risk Guide (ICRG), and Political Instability (formerly, State Failure) Task Force (PITF) datasets.

The eight variables from the ICRG that we include in our EFA are:

government stability (0–12), which assesses "the government's ability to carry out its declared programs and ability to stay in office," based on three subcomponents, Government Unity, Legislative Strength and Popular Support, each with a maximum score of four points (very low risk) and a minimum score of zero points (very high risk);

The *democratic accountability index* (0–6) which measures how responsive a government is to its people by tracking the system of government (for example, a system with a varied and active opposition is assigned a higher score than one where such opposition is limited or restricted);

The *investment profile* index (0–12), which captures the enforcement of contractual agreements and expropriation risk (countries with lower risk are higher in the index);

The *corruption index* (0–6), which measures the *absence* of the kinds of corruption, such as nepotism, bribes, etc. that if revealed, may lead to political instability such as the overthrow of the government or the breakdown of law and order; the index of *bureaucratic quality* (0–4), which assesses the efficiency and autonomy of the bureaucracy;

Internal conflict, which captures the *absence* of internal civil war;

External conflict, which similarly measures the absence of foreign wars; and

Ethnic tensions, which provides an *inverse* measure of the extent to which racial and ethnic divisions lead to hostility and violence.

Next, we include nine variables from the DPI dataset. They are: *legislative fractionalization*, which captures how politically diverse a system is by looking at the number of parties participating in a regime; *political polarization* (0–2) measures the ideological distance between the legislature and the executive; *Executive years in office*, the number of years the current chief executive has served; *Changes in veto power* which measures the percent drop in the number of players who have veto powers in the government (if president gains control of the legislature, veto power drops from 3 to 1); a *Government Herfindahl-Hirschman index* that measures the degree to which different parties share in the operation of the government, measured as the sum of the squared seat shares of all parties in the government; *the number of veto players within the government*; whether *allegations* of fraud, boycott by important opposition parties, or candidate intimidation, surfaced in the last election (less fraud = higher rank); the *legislative index of electoral competition* (1–7) which measures the degree to which the selection of the legislature is decided by elections (no legislature = 1); and the *executive index of electoral competition* (1–7) which measures the degree to which the selection of the executive is decided by free and fair elections (executive elected directly by people = 7).

To these we add nine measures from the CNTS, which are: *Assassinations* (the number of times there is an attempt to murder, or an actual murder of, any important government official or politician), *Strikes* (the number of times there were mass strikes by 1000 or more industrial or service workers, across more than one employer, protesting a government policy), *Government Crises* (the number of times there was a crisis that threatened the downfall of the government, other than a revolt specifically to that end), *Demonstrations* (the number of times there was peaceful protestations of government domestic policy by 100 or more people), *Purges* (the number of times political opponents, whether part of the regime, or outside it, were systematically eliminated), *Riots* (the number of times there was a violent protest by 100 or more people), *Cabinet changes* (the number of times either a new premier was named and/or the number of times new ministers replaced 50% of the cabinet positions with fewer changes indicating a more stable government), *Change in Executive*[1] (the number of times in a year that effective control of executive power went to a new independent chief executive), and the *Legislative Effectiveness Index*, 0–3 (measures how independent the legislative is of the executive, and therefore how effective it is, 0 = no legislature).

Finally, we include four variables from the PITF: The Polity 2 Democracy index (measures the institutional regime, ranging from −10, institutional autocracy to +10, institutional democracy), regime durability (the number of years since the most recent change in regime or the end of a period of politically unstable institutions, the more stable the region, the higher the number), ethnic wars, and nonethnic civil war.

---

[1]Executive (0–3): Coded as following, 1 for direct election, 2 when the election is Indirect, and 3 if it is considered a nonelective. Direct Election is when the election of the effective executive is by popular vote or by delegates committed to executive selection. Indirect Election is when the chief executive is elected by an elected assembly or by an elected but uncommitted Electoral College or when a legislature is called upon to make the selection in a plurality situation. Nonelective is when the chief executive is chosen neither by a direct or indirect mandate.

2.2   Problems with Institutional Measures

Simply including a subset of those measures is problematic for three reasons. First, although they purport to gauge distinct aspects of institutional character, many of them overlap substantially, and most of them are highly correlated with one another. Second, the subjective nature of these de facto indices of quality may expose them to considerable measurement error. Third, institutional quality has been shown to be multi-dimensional (Bang et al. 2017), and the different dimensions may have different impacts. The nonparametric methodology that we adopt partially avoids that issue in the sense that we do not need to worry about obtaining biased parameter estimates. However, we do need to worry that similar measures of institutional quality that represent the same underlying concept might dominate our classification.

In order to purge our institutional measures from some of these problems we perform an exploratory factor analysis (EFA) on the institutional measures described above. EFA is similar in some respects to the more familiar technique of principle components analysis (PCA) in that both EFA and PCA reduce the dimensionality of the observed variables based on the variance-covariance matrix. However, in contrast to PCA, which seeks to extract the *maximum* amount of variation in the correlated variables, EFA seeks to extract the *common* sources of variation. To achieve this, EFA expresses the observed variables as linear combinations of the latent underlying factors (and measurement errors), whereas PCA expresses the latent components as combinations of the observed variables.

We report the results of the factor analysis Table 2.4 below. From the factor loadings, we identify seven common factors out of the list of institutional variables:

Democracy is comprised by the Polity index; the legislative and executive indices of electoral competition; legislative fractionalization; and democratic accountability.

Violence consists of the internal and external conflict indices, ethnic tensions, and the presence of ethnic conflict and civil war. Higher scores indicate greater *stability*.

Transparency incorporates the corruption, bureaucratic quality, and democratic accountability indices, along with regime durability and fraud.

Protest is constructed primarily from the numbers of demonstrations, riots, and strikes in society. Higher numbers indicate greater *unrest*.

Within-Regime Instability includes legislative concentration and fractionalization, as well as political polarization. Countries with more fractious governments receive higher values.

Credibility is formed by the investment profile and government stability indices.

Regime Instability is composed of the numbers of executive changes and major cabinet changes, along with the changes in veto players and executive tenure.

One useful feature of these results is that they bear a striking similarity to the factors previously derived by Jong-a-Pin (2009) and Bang and Mitra (2011). For this reason, we have applied the same terms in our interpretation of these factors. In this sense, our results are quite consistent with previous contributions to the literature on institutions that employ factor analysis. Last, EFA's control for unsystematic measurement error. To this extent, they help contribute to resolving the measurement error problem rife in the growth literature.

Table 2.4 EFA results

Democracy	5.906				Observations	10,806
Violence	4.543				Retained factors	8
Protest	1.301				Parameters	212
Regime instab.	1.176				LR stat	173,460
Transparency	0.885				P-value	0.000
Within-regime	0.821					
Credibility	0.465					
Factor8	0.465					

	Democracy	Violence	Protest	Regime	Transparency	Within	Credibility	Factor8	Uniqueness
Leg. frac.	**0.777**	0.006	0.003	0.085	−0.084	0.430	0.029	0.025	0.195
Pol. polariz.	**0.613**	0.146	0.010	−0.005	0.166	0.344	−0.048	−0.033	0.454
Exec. tenure	−0.473	0.064	0.028	−0.340	−0.218	0.006	−0.127	−0.093	0.583
Vetoes	0.131	−0.074	−0.012	0.300	0.131	−0.008	0.083	0.073	0.858
Gov. herf.	−0.403	0.009	0.055	−0.047	−0.017	**−0.660**	−0.016	0.010	0.395
Checks	**0.820**	0.087	0.020	0.010	0.038	0.123	−0.027	−0.045	0.300
Leg. elec. com.	**0.857**	0.120	0.027	−0.060	−0.210	−0.119	−0.039	−0.040	0.185
Exec. elec. com.	**0.916**	0.111	0.008	−0.017	−0.147	−0.142	−0.012	0.016	0.105
Fraud	−0.128	−0.235	−0.032	−0.029	−0.377	0.010	−0.041	0.104	0.772
Polity2	**0.860**	0.113	0.008	0.050	0.167	−0.018	0.049	0.020	0.213
Reg. dur.	0.200	0.241	0.107	−0.274	0.324	−0.015	0.044	−0.054	0.705
Eth. conf.	−0.027	**−0.625**	0.065	−0.025	0.090	0.087	0.228	−0.184	0.502
Civil war	−0.029	−0.445	0.060	0.006	0.023	−0.011	−0.000	0.483	0.563
Assassin	0.059	−0.198	0.173	0.089	−0.039	−0.021	−0.007	0.249	0.855
Strikes	0.089	−0.110	0.407	0.120	−0.003	−0.006	−0.112	0.029	0.786
Gov. crises	0.108	−0.125	0.227	0.301	−0.032	0.055	−0.106	0.065	0.811
Purges	−0.058	−0.038	0.182	0.133	−0.078	0.012	−0.065	0.033	0.933

Riots	0.060	−0.090	**0.708**	0.033	0.008	−0.020	0.011	0.001	0.485
Demonstrations	0.052	−0.088	**0.695**	−0.007	0.018	−0.024	0.007	0.013	0.506
Cab. changes	0.073	−0.195	0.052	**0.579**	−0.120	0.056	−0.041	−0.013	0.599
Exec. changes	0.243	−0.061	0.039	**0.587**	0.049	0.041	−0.028	−0.016	0.586
Leg. elections	0.334	0.099	0.026	0.175	0.027	−0.076	−0.071	−0.080	0.830
Gov. stab.	0.027	**0.673**	−0.036	−0.129	−0.136	0.046	0.372	0.009	0.369
Dem. acct.	**0.737**	0.446	0.000	0.051	0.216	−0.006	0.051	0.020	0.206
Invest. profile	0.290	**0.716**	−0.019	−0.095	0.012	0.030	0.380	0.000	0.249
Corruption	0.480	0.486	0.014	−0.028	0.462	0.020	−0.145	0.083	0.290
Bur. qual.	**0.503**	**0.605**	0.045	−0.067	0.357	0.051	0.065	−0.022	0.240
Eth. tensions	0.119	**0.743**	−0.026	0.010	−0.004	−0.039	−0.148	0.212	0.364
Int. conflict	0.201	**0.898**	−0.045	0.007	0.018	0.001	−0.077	−0.190	0.109
Ext. conflict	0.322	**0.708**	0.001	0.049	−0.040	−0.000	−0.062	−0.014	0.387

The bold highlights the components that have higher loads in each factor and therefore define the factor

2.3   Imputing Missing Data

Another problem with many empirical studies of growth is that many of the variables are missing for a substantial portion of any time sample. Therefore, simply cobbling together a dataset that includes a diverse range of input variables *and* covers a wide range of countries over a long period is nearly impossible. A secondary consequence, therefore, is that any study of growth must trade off bias resulting from sample selection on the one hand, against omitted variables on the other hand.

Tree-based Machine Learning techniques deal with the problem well because if data for the optimal splitting variable at any particular node is missing for an observation, the algorithm can substitute the missing information in one of two ways. First, a regression tree will attempt to complete the splits using surrogate information from other variables that track the values of the optimal splitting variables very closely. If that is not possible, then the tree model will split the missing values based on the conditional median (or mode for categorical variables) for the observations in that node.

Thus, Machine Learning actually suggests a useful way to impute data: Replace missing values in the dataset with the median (mode) value, conditional on the observed values of both the target and input variables up until reaching the node where the model encountered the missing values. While this imputation tactic may not be ideal for a single iteration of a tree model, conditioning the imputed values on the observed inputs and outputs of a few *hundred* random trees (as would be the case with the Random Forest model) is likely to yield reasonably good imputed values. Studies that have tested the validity of Random Forest imputation using simulated missing values have found that this imputation method performs comparably, and often better than, other methods of imputation (such as multiple imputation and OLS).

Takeaways

1. We include data that mirrors Sala-i-Martin's (1997) list of robust covariates from EBA analysis. These variables originate from the major strands of theories of growth. Therefore, they represent major growth theories quite comprehensively.
2. Some of the variables can be deconstructed to non-overlapping dimensions using EFA. This process also controls for unsystematic measurement errors.
3. Machine Learning fills in missing data with validated imputation techniques.

References

Bang, J. T., & Mitra, A. (2011). Brain drain and institutions of governance: Educational attainment of immigrants to the US 1988–1998. *Economic Systems, Elsevier, 35*(3), 335–354.

Bang, J. T., Basuchoudhary, A., & Mitra, A. (2017, April 1). The machine learning political indicators dataset. Retrieved from Researchgate: https://www.researchgate.net/publication/316118794_The_Machine_Learning_Political_Indicators_Dataset

Jong-A-Pin, R. (2009). On the measurement of political instability and its impact on economic growth. *European Journal of Political Economy, 25*(1), 15–29.

Sala-i-Martin, X. (1997). I just ran four million regressions. *American Economic Review, 87*, 178–183.

Solt, F. (2016). The standardized world income inequality database*. *Social Science Quarterly, 97* (5), 1267–1281.

Chapter 3
Methodology

We build an empirical model using Machine Learning techniques (Artificial Neural Network, Boosting, Bootstrap Aggregating, Random Forest predictors and Regression Tree) to provide an objective approach to finding *linear and non-linear* patterns on *how* publicly-available economic, geographic, and institutional variables *predict* growth (Hand et al. 2001). First, we identify the Machine Learning approach that best predicts growth. Then, using the best technique, we identify the variables that form the pattern that best predicts growth. This is important because, at least theoretically, there may be reason to believe that many of the correlates of growth have complex non-linear impacts of growth because strategic complementarities between variables can lead to multiple possible growth equilibria. For example, openness to trade may improve growth *on average*, but only if a country is not too dependent on natural resources or other primary commodities that may lead to conflict among competing factions over the rents generated by the resource sector.

Machine Learning techniques identify tipping points in the range of a particular variable that may place a country in a lower or higher growth category. Moreover, Machine Learning can generate partial dependence plots. These graphs can illustrate how variables identified as good predictors of growth relates (perhaps non-linearly) to growth. Further, by identifying the variables that have the *most* predictive power we could help develop a framework to distinguish between competing theoretical explanations of growth. Suppose, for instance, political economy models may suggest that income inequality is important in explaining growth, but neoclassical models may predict that education matters more. If Machine Learning methodologies rank income inequality as a better predictor of growth than population density, we can assume that the political economy model may itself be a better explanation of growth than the neoclassical model, or vice versa. This would then suggest greater econometric scrutiny of Theory A in teasing out causal patterns. Moreover, this Machine Learning approach can help eliminate correlates of conflict that do not predict economic growth well. Presumably, correlates that do not predict well cannot really be considered as variables that cause growth.

© The Author(s) 2017   19
A. Basuchoudhary et al., *Machine-learning Techniques in Economics*,
SpringerBriefs in Economics, https://doi.org/10.1007/978-3-319-69014-8_3

Our Machine Learning approach, and the econometric tests arising out of this approach, will help us better understand causal patterns explaining growth. Moreover, we offer a better understanding of how growth can be *predicted*, which will be of particular help to policy makers as they design policies of economic growth. Below, we outline and explain the different estimation techniques that we use.

3.1 Estimation Techniques

In general, using given data from a learning sample, $L = \{(y_1, \mathbf{x}_1), \ldots (y_N, \mathbf{x}_N)\}$, any prediction function, $d(\mathbf{x}_i)$, maps the vector of input variables, $\mathbf{x}$, into the output variable, y. An effective prediction algorithm seeks to define parameters that minimize some error function over the predictions. Common error functions that many predictors use include the mean (or sum) of the absolute deviations of the observed values from the predicted values or the mean (or sum) of the squared deviations. In linear regression models $d(\mathbf{x}_i)$ is simply a linear function of the inputs and their respective slope coefficients, plus a constant, $d(\mathbf{x}_i) = \mathbf{x}_i\beta$. For linear models, we can express the minimization condition as:

$$R_{MAD}(d) = \frac{1}{n} \sum_{i=1}^{N} |y_i - d(\mathbf{x}_i)|,$$

And:

$$R_{OLS}(d) = \frac{1}{n} \sum_{i=1}^{N} (y_i - d(\mathbf{x}_i))^2,$$

Where $d(\mathbf{x}_i) = \mathbf{x}_i\beta$ is a linear function of the inputs.

Under certain conditions, the predictor $(\mathbf{x}_i\beta)$ that minimizes the mean absolute deviations function can be shown to be a good estimate for the conditional *median* (or other conditional quantiles if the absolute deviation function is "tilted" as in quantile regression). Correspondingly, the predictor that minimizes the least squares function can be shown to be a good estimate for the conditional *mean*, or expected value of y, $E(y|\mathbf{x})$.

Although linear regression can sometimes yield good predictors, it is important to realize that the main objective of linear models is to estimate *causal* effects for one or more hypothesized determinants of y holding all of the other hypothesized determinants in a model constant. More sophisticated methods of estimating linear regression models (such as ones that use instrumental variables or other two-step methods) focus on purging the marginal causal effects of bias that might result from endogeneity, selection bias, or misspecifications of the functional form of the target variable. These methods address the problem of bias in the estimated marginal effects to the detriment of the model's overall predictive accuracy.[1]

Other modifications of the Classical Regression Model include techniques based on *maximum likelihood estimation* (MLE). Under the classical linear assumptions (including the assumption that the errors are *i.i.d.* normally distributed), MLE and least squares estimation will be equivalent. When the errors are not normal, such as is the case with a binomial (or ordinal or multinomial) dependent variable or count variables, MLE methods (logit and negative binomial regression, for example) are needed to account for the breach of the assumption.

3.1.1 Artificial Neural Networks

Artificial neural networks (ANNs) allow us to increase the complexity of our prediction algorithm by chaining together a set of linear (or quasi-linear in the case of logit) ones. A diagram of a simple, two-layer ANN with five inputs and two hidden units appears below:

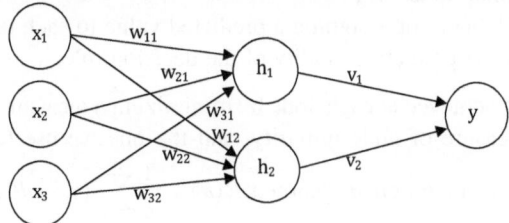

In the diagram, the inputs, x_1, x_2, and x_3, are the observed explanatory variables: the hidden units, h_1 and h_2, mix the variables before they are re-mixed to predict the output, y, in the second layer. The parameters in the model consist of the two layers of trained weights: w_{ij} are the weights that connect input x_i to hidden layer h_j; v_j are the weights that connect hidden layer h_j to the output y.

One detail that we have not shown in the diagram is the *link function* (or activation function) that maps the linear interactions of variables to the output space. For prediction, it is not unreasonable to use a simple linear link function, whereas for classification it is more common to use a logistic link function or a

---

[1]For example, it is fairly well-known that a two-stage least squares estimator can sometimes yield a *negative* value for the regression R^2. This implies that the sum of squared errors of the model exceed the total sum of squares of the target variable.

hyperbolic tangent (*tanh*) function, both of which map the real line to the range (0,1). In our neural networks, we use allow 20 units in the hidden layer, and linear and logistic link functions for the growth regression and recession classification problems, respectively.

3.1.2 Regression Tree Predictors

Classification and regression trees (CART)[2] diagnose and predict outcomes by finding binary splits in the input variables in order to optimally divide the sample into subsamples with successively higher levels of purity in the outcome variable, y. So, unlike linear models, where the parameters are linear coefficients on each predictor, the parameters of the tree models are "if-then" statements that split the dataset according to the observed values of the inputs.

More specifically, a tree, T, has four main parts:

1. Binary splits to splits in the inputs that divide the subsample at each node, t;
2. Criteria for splitting each node into additional "child" nodes, or including it in the set of terminal nodes, T^*;
3. A decision rule, $d(\mathbf{x})$, for assigning a predicted value to each terminal node;
4. An estimate of the predictive quality of the decision rule, d.

The first step is achieved at each node by minimizing a measure of impurity. The most common measure of node impurity, and the one we use for our tree algorithms, is the mean square error, denoted $\hat{R}(d) = \frac{1}{n} \sum_{i=1}^{N} (y_i - d(\mathbf{x}_i))^2$. Intuitively, this method searches for the "best" cutoff in each of the inputs to minimize errors, then selects which of the inputs yields the greatest improvement in node impurity, using its optimal splitting point. Then, a node is declared to be terminal in the second step if one of the following conditions is met: (1) that the best split determined by the application of step fails to improve the node impurity by more than a predetermined minimum improvement criterion; or (2) the split creates a "child" node that contains fewer observations than the minimum allowed.[3] At each

---

[2]We provide only a brief summary of tree construction as it pertains to our objectives. For a full description of the CART algorithm, see Breiman et al. (1984).

[3]Note that there is a tradeoff here: setting lower values for the minimum acceptable margin of improvement or the minimum number of observations in a child node will lead to a more accurate predictor (at least within the sample the model uses to learn). However, improving the accuracy of the predictor within the sample will also lead to a more complex (and therefore less easily-interpreted) tree, and may lead to over-fitting in the sense that the model will perform more poorly out-of-sample.

terminal node, the decision rule typically assigns observations with a predicted outcome based on the outcome that is most frequent (more than one-half of the observations in that node for binary outcomes, for example).[4]

The predictive quality of the rule is also evaluated using the *mean square error*, $\hat{R}(d) = \frac{1}{n} \sum_{i=1}^{N} (y_i - d(x_i))^2$. This misclassification rate is often cross-validated by splitting the sample several times and re-estimating the misclassification rate each time to get an average misclassification of all of the cross-validated trees.

3.1.3   Boosting Algorithms

Combining ensembles of trees can often improve the predictive accuracy of a CART classifier. The bootstrap aggregating (bagging) predictor, boosting (adaptive boosting and other generalizations of boosting) algorithm, and random forest predictors all predict outcomes using ensembles of classification trees. The basic idea of these predictors is to improve the predictive strength of a "weak learner" by iterating the tree algorithm many times by either modifying the distribution (boosting) or randomly resampling the distribution (bagging). Then either classify the outcomes according to the outcome of the "strongest" learner once the algorithm achieves the desired error rate (boosting), or according to the outcome of a vote by the many trees (bagging).

Boosting is a way proposed by Freund and Schapire (1997) to augment the strength of a "weak learner" (a learning algorithm that predicts poorly) and making it a "strong learner." More specifically, for a given distribution $\mathcal{D}$ of weights assigned to each observation in L, and for a given desired error $\tilde{R}$, and failure probability, ϕ, a *strong learner* is an algorithm that has a sufficiently high probability (at least $1-\phi$) of achieving an error rate no higher than $\tilde{R}$. A weak learner has a lower probability (less than $1-\phi$) of achieving the desired error rate. Adaboost creates a set of M classifiers, $F = (f_1, \ldots, f_M)$ that progressively re-weight each observation based on whether the previous classifier predicted it correctly or incorrectly. Freund and Schapire (1997) and Friedman (2001) have developed modifications of the boosting algorithm for classification for regression trees.

Starting with a $\mathcal{D}_1 = (1/N, \ldots, 1/N)$, suppose that our initial classifier, $f_1 = T$ (the Single Tree CART predictor) is a "weak learner" in that the misclassification rate, $\hat{R}(d)$ is greater than the desired maximum desired misclassification rate, $\tilde{R}$. Next,

---

[4]It is possible, however, to consider decision rules that assign one class of a binary outcome anytime the proportion of observations exceeds one-third of the total observations in that node, especially if one type of misclassification error is costlier than the other.

for all observations in the learning sample, recalculate the distribution weights for
the observations as:

$$D_2 = \frac{\mathcal{D}_1(i)}{Z_2} \times \begin{cases} \dfrac{\hat{R}_1(d)}{1 - \hat{R}_1(d)} & \text{if } d_1(\mathbf{x}_i) = y_i \\ 1 & \text{otherwise} \end{cases},$$

Where Z_m is a scaling constant that forces the weights to sum to one.

The final decision rule for the boosting algorithm is to categorize the outcomes
according to $d(x) = \arg\max_{y \in Y} \sum_{m:d_m(\mathbf{x})=y} \log\left(\dfrac{1 - \hat{R}_m(d)}{\hat{R}_m(d)}\right)$. Using this decision rule
and its corresponding predictions, we calculate the estimate of the misclassification
rate in the same way as in step (4) of the Single Tree algorithm.

3.1.4   Bootstrap Aggregating (Bagging) Predictor

The bagging predictor proposed by Breiman (1996) takes random resamples $\{L^{(M)}\}$
from the learning sample *with replacement* to create M samples using only the
observations from the learning sample. Each of these samples will contain
N observations—the same as the number of observations in the full training sample.
However, in any one bootstrapped sample, some observations may appear twice
(or more), others not at all.[5] The bagging predictor then adopts the rules for splitting
and declaring nodes to be terminal described in the previous section to build
M classification trees.

To complete steps (3) and (4), the bagging predictor needs a way of aggregating
the information of the predictions from each of the trees. The way that bagging
predictors (and other ensemble methods) do this for class variables is through
voting. For *classification trees* (categorical target variables), the voting processes
each observation[6] through all of the M trees that was constructed from each of the
bootstrapped samples to obtain that observation's predicted class for each tree. The
predicted class for the entire model, then, is equal to the mode prediction of all of
the trees. For *regression trees* (continuous target variables), the voting process
calculates the mean of the predicted values for all of the bootstrapped trees. Finally,

---

[5]Note that the probability that a single observation is selected in each draw from the learning set is
$1/N$ Hence, sampling with replacement, the probability that it is completely left out of any given
bootstrap sample is $(1 - 1/N)^N$. For large samples this tends to $1/e$. The probability that an
observation will be completely left out of all M bootstrap samples, then, is $(1 - 1/N)^{NM}$.

[6]Note that the observations under consideration could be from the in-sample learning set or from
outside the sample (the test set).

the predictor calculates the redistribution estimate in the same way as it did for the single classification tree, using the predicted class based on the voting outcome for each predictor.

3.1.5 Random Forests

Like the bagging predictor, the random forest predictor is a tree-based algorithm that uses a voting rule to determine the predicted class of each observation. However, whereas the bagging predictor randomizes the selection of the observations into the sample for each tree, and then builds the tree using the same procedure as CART, the random forest predictor may randomize over multiple dimensions of the classifier (Breiman 2001). The most common dimensions for randomizing the trees are the selection of the inputs for node of each tree, as well as the observations included for constructing each of the trees. We briefly describe the construction of the trees for the random forest ensemble below.

A random forest is a collection of tree decision rules, $\{d(x), \Theta_m = 1, \ldots, M\}$, where Θ_m is a random vector specifying the observations and inputs that are included at each step of the construction of the decision rule for that tree. To construct a tree, the random forest algorithm takes to following steps:

 i. Randomly select $n \leq N$ observations from the learning sample[7];
 ii. At the "root" node of the tree, select $k \in K$ inputs from $\mathbf{x}$;
 iii. Find the split in each variable selected in (ii) that minimizes the mean square error at that node and select the variable/split that achieves the minimal error;
 iv. Repeat the random selection of inputs and optimal splits in (ii) and (iii) until some stopping criteria (minimum improvement, minimum number of observations, or maximum number of levels) is met.

The bagging predictor described in the previous sub-section is in fact a special case of the random forest estimator where, for each tree, Θ_m consists of a random selection of $n = N$ observations from the learning sample with replacement (and each observation having a probability of being selected in each draw equal to $1/N$) and sets the number of inputs to select at each node, k, equal to the full length of the input vector, K so that all of the variables are considered at each node.

---

[7]In contrast to bagging, where the number of observations selected for each tree exactly equals the total number of observations in the learning sample, and the draws always sampled with replacement, the number of observations selected for each tree of the forest can be set to be less than the total size of the learning sample, and can therefore be sampled with *or without* replacement. This also allows for slightly greater flexibility with respect to stratified or clustered sampling from the learning sample.

3.2   Predictive Accuracy

Once we have built our dataset and imputed the missing values, we evaluate the validity of our error estimates and the predictive strength of our models. Error estimates ($R[d]$) can sometimes be misleading if the model we are evaluating is over-fitted to the learning sample. These error estimates can be tested out-of-sample and also cross-validated using the learning sample.

To test the out-of-sample validity, we simply split the full dataset into two random subsets of *countries*: the first, known as the *learning sample* (or training sample) contains the countries and observations that will build the models; the second, known as the *test sample*, will test the out-of-sample predictive accuracy of the models. The out-of-sample error rates will indicate which models and specifications perform best, and will help reveal if any of the models are over-fitted.

For the growth models, we calculate the redistribution estimate as the MSE. For the recession classification models, we calculate the redistribution estimate as the proportion of the sample predicted incorrectly. In addition to the standard redistribution estimate, other measures of accuracy can sometimes be useful. Therefore, in addition to the redistribution (or overall error) rate, we calculate the positive predictive value (PPV = true positives/predicted positives), negative predictive value (NPV = true negatives/predicted negatives), sensitivity (true positives/ *observed* positives), and specificity (true negatives/*observed* negatives). Table 3.1 summarizes the calculations of these additional measures:

We also report the area under the receiver-operating characteristic (ROC) curve (AUC). The ROC curve plots the sensitivity (true positive rate) on the vertical axis against one minus the specificity (false positive rate) on the horizontal axis. Each point on the ROC curve corresponds to the sensitivity and specificity for a different threshold probability for classifying the predicted class as "positive" (recession). As this threshold probability increases, the model predicts positive outcomes with higher probability, and therefore the sensitivity increases; however, this will increase the number of observed negatives predicted to be positive and decrease the specificity (increase the false positive rate). The area under this curve in some sense can represent a measure of how well the model balances these types of accuracy.

Note that a high AUC (or any other single measure of accuracy for that matter) in the *learning sample* is not always indicative of a strong learner. To give an extreme example, suppose that, at a standard threshold of 0.5, a model predicts exactly *zero* positive outcomes. In this case, the *specificity* of the model would be *1*—the best

Table 3.1 Summary of how different error measures are calculated

		Observed outcome		
	Total sample	Positive	Negative	
Predicted outcome	Positive	True positive	False positive	PPV
	Negative	False negative	True negative	NPV
		Sensitivity	Specificity	Overall error

possible value. However, the same learner would have a sensitivity and positive predictive value of zero. Next, suppose that all of the predicted probabilities for the learning sample are very tightly distributed, let's say around 0.25 ± 0.01, but that the probabilities *order* the outcomes *perfectly* in the sense that all of the negative outcomes have the lowest probabilities, and all of the positive outcomes have the highest probabilities. In this case, there will be an "optimal" threshold (in the sense of maximizing the combined sensitivity and specificity) for the learning sample that will imply a perfect predictor. However, it is highly unlikely that the test sample will line up in a perfect order around that same threshold, and therefore the model will probably predict very poorly in the test sample. In this case, the model will be *over-fitted* to the learning sample.

As a final note, when sample sizes are relatively small, machine learning can use a method known as cross-validation to validate the predictive accuracy. Cross-validation creates V random sub-samples of the data, predicts the observations in those sub-samples using the full model, and calculates the error estimate for each of them. It then averages the error estimates of all V random subsamples to give the cross-validated error estimate. Small samples are not a problem for our dataset, and so we use a fully separate test sample instead of cross-validation.

3.3 Variable Importance and Partial Dependence

We can also use tree models and tree ensembles to assess model selection and the approximate direction and functional form of the relationships between the inputs and the output. Here, we discuss measures of *variable importance* in tree models, and then describe *partial dependence plots* as a tool for describing the direction and functional form of the impacts.

Regression trees measure a variable's importance based on the margin by which the variable reduces the MSE of the model. It does this by adding up the reduction in the MSE at each node where that variable determines a split. Ensemble methods achieve this measure in a similar way, except that the importance of each variable takes the average of the importances from the nodes in all of the trees that select that variable. For each model, we standardize the importance measure by dividing the decrease in the MSE that the model attributes to each variable by the total reduction in the MSE for the model as a whole.

Classification trees produce a couple of different options: the average decrease in the overall node impurity (which is similar to the MSE), and the average decrease in the overall Gini dispersion. In our models, we have chosen to use the average decrease in Gini dispersion because all of the classification algorithms calculate this measure as a part of the standard list of output values. We also standardize the variable importances so that we can better judge the contribution of each variable relative to the other variables across models.

Another useful tool for machine learning models is a Partial Dependence Plot (PDP). PDPs display the marginal effect of variable x_k conditional on the observed values of all of the other variables, $(x_{1,-k}, x_{2,-k}, \ldots x_{n,-k})$. Specifically, it plots the graph of the function:

$$\hat{f}(x) = \frac{1}{n} \sum_{i=1}^{n} f(x_k, x_{i,-k}),$$

Where the summand, $f(x_k, x_{i,-k})$, is defined by:

$$f(x) = \ln[p_1(x)] - \frac{1}{C} \sum_{s=1}^{C} \ln[p_s(x)]$$

For classification models and by the value of the target variable (i.e. the growth rate) for regression models. In the equation, s indexes the classes of the target variable and C is the total number of classes. So, if an increase in a given factor, x_k, increases the probability of recession ($s = 1$), then the value of the function plotted by the PDP also increases.

PDPs are useful in machine learning models because they can help to give an idea of the magnitude and direction of the impacts of the predictors. In addition, by adding tick marks corresponding to the decile cutoffs for the distribution of the input variable being plotted, one can get an idea of how the target variable varies across the distribution of each input (explanatory) variable.

References

Breiman, L. (1996). Bagging predictors. *Machine Learning, 24*(2), 123–140.
Breiman, L. (2001). Random forests. *Machine Learning, 45*(1), 5–32.
Breiman, L., Friedman, J., Stone, C. J., & Olshen, R. A. (1984). *Classification and regression trees*. Boca Raton, FL: CRC Press.
Freund, Y., & Schapire, R. E. (1997). A decision-theoretic generalization of on-line learning and an application to boosting. *Journal of Computer and System Sciences, 55*, 119–139.
Friedman, J. (2001). Greedy function approximation: A gradient boosting machine. *Annals of Statistics, 29*, 1189–1232.
Hand, D., Mannila, H., & Smyth, P. (2001). *Principles of data mining*. Cambridge: MIT Press.

Chapter 4
Predicting a Country's Growth: A First Look

We start this chapter with some stylized facts about economic growth. First, there are wide disparities in global income even after converting the per capita GDP of countries to reflect Purchasing Power Parity (PPP) (see for example the Penn World Tables, Summers and Heston 1988; the most recent update is available here: https://fred.stlouisfed.org/categories/33402). Second, despite these inequalities, over long periods of time, most countries have experienced positive growth in their GDP. Further, there are growth miracles and disasters—North and South Korea immediately spring to mind (though North Korean data is suspect). Some countries, e.g. some in Sub Saharan Africa, appear to have maintained their positions in the global distribution of income, while others, like China have begun to catch up to the world's richer countries. In short, almost anything seems possible as far as economic growth is concerned. However, a brief perusal of any macroeconomic textbook will reveal chapters on growth models that start with a Solow type model and possibly move on to the Romer type of endogenous growth models. The textbook approach suggests a consensus in understanding economic growth as some decoction of capital (both human and physical) and labor. The wide variation in the world's experience of growth across different countries, however, suggests that the textbook approach to growth is not in sync with our experience.

We report results of our efforts on predicting economic growth in this chapter. At the outset, we should emphasize our disappointment in our ability to predict economic growth. As we noted in the introduction, we do a much better job in predicting recessions than predicting growth. Nevertheless, our problems with prediction expose some of the limitations of theoretical growth models in a predictable way. We suggest that our results here expose some of the limitations of popular econometric investigative strategies applied in the growth literature as well. These results may be controversial. We however, merely suggest that we now have the computing technology to start questioning some of the orthodoxy in economic growth as a matter of validated science. Taken together we believe this chapter has some important implications for theoretical growth models, econometric evidence thereof, and possible policy implications as well.

© The Author(s) 2017

A. Basuchoudhary et al., *Machine-learning Techniques in Economics*,
SpringerBriefs in Economics, https://doi.org/10.1007/978-3-319-69014-8_4

Table 4.1 Comparing predictive quality

	Learning sample		Test sample	
	MSE	% Reduction	MSE	% Reduction
OLS regression	10.423061	0.281875832	14.96455	−0.015089438
Neural network	14.469619	0.003077588	15.0638	−0.021821857
Single tree	4.763529	0.671804156	26.02018	−0.765025336
Bagging	6.769989	0.533563823	15.54106	−0.054195807
Boosting	12.694542	0.125376181	14.11881	0.042279594
Random forest	6.505755	0.551768919	12.94524	0.121886298
Average prediction	8.231043	0.43290067	12.87478	0.126665807

We report our findings of the predictive quality of our different algorithms, Artificial Neural Networks, Decision Trees, Bagging, Boosting, OLS Regression and Random Forests in Table 4.1. This endeavor takes the following form.[1] First, we randomly divide our dataset into two parts—a learning part and a validation or test sample part. The algorithms are "trained" in the learning sample, and such training is complete when the algorithm has minimized the mean square error (MSE) within the sample. Each algorithm follows a similar process of minimizing the mean square error in the learning sample. We then check how this model performs in minimizing the mean square error in the test sample.[2] How well an algorithm, one that has done the best possible job in minimizing the mean square error in the training sample, reduces error in a sample *it has never seen before* is a hard test of predictive quality. Algorithms that fit the test sample better in terms of lower mean square errors are more predictive than others. We report the mean square errors in both the learning and the test samples in Table 4.1 for six different techniques (Logit, Single Classification tree, Bagging, Random Forest,[3] Adaboost,[4] and ANN). We also report how much each algorithm changes the mean square error relative to a baseline total MSE with zero covariates (mean sum of squared deviations from the unconditional mean) to get a sense of how well a model might predict relative to a simple average.[5] We do find algorithms that minimize the mean square error.

---

[1] What follows is a summary of the process we detail in the methodology chapter to help the reader recall some of the methodological fundamentals.

[2] We chose mean square error (MSE) as a fairly straightforward measure for comparing predictive errors since the economic growth is a continuous variable.

[3] Random forest predictors are also multi-tree predictors, except instead of randomizing the observations for each tree, it randomizes the variables selected at each node of each tree.

[4] The Boosting predictor is yet another multiple-tree predictor. It builds each tree by successively increasing the weight on misclassified observations in order to reduce the overall misclassification rate in the learning set.

[5] Total or Unconditional MSE: $\sum (Y_i - E(Y))^2/n$. Model or Conditional MSE: $\sum (Y_i - E(Y|X))^2/n$. The percent reduction is the (Total − Model)/Total.

We note first that the Random Forest algorithm provides the best predictive quality in that it has the lowest mean square error in both the learning and the test sample. The MSE for the Random Forest method in the learning sample is in 6.5 while in the test sample it is 12.95. However, the percentage reduction in MSE due to the algorithm is a mere 0.12%. One can get a sense of the poverty of prediction by looking at the root mean square error in the test sample. The prediction error can be as large as 3.6%. When it comes to predicting growth rates, that is a big number. The other algorithm methodologies do much worse in terms of mean square error and in some cases, as in the OLS, ANN, Single Tree, and Bagging methods, they increase the predictive error relative to the baseline total MSE. So, what might all this mean in the context of understanding economic growth?

Recall that our predictive models use variables identified by Sala-i-Martin (1997) as the universal correlates of growth. These variables are not chosen at random. Economic growth theory models undergird Sala-i-Martin's variable choice. Arguably, a "good" theoretical model of economic growth should be predictive. After all, a good theory of growth must explain the reality of growth everywhere and not just within the confines of one sample dataset. There seems to be a limitation of theoretical growth models in this regard because the variables chosen to represent the theoretical models do not do seem to be well validated *out-of-sample* in the sense that the predictive error seems to be quite large.

The growth literature itself documents many reasons for this. For example, systematic measurement error may be the reason for the weak predictive power of our model (Temple 1999). Most studies usually ignore these errors. We correct the sources of the most egregious errors (presumably in the subjective institutional indices) through our Exploratory Factor Analysis. Beyond this, our caveat would be the same as any other growth study given that we choose variable sources that are well established in the field. Nevertheless, since both the learning and test samples are from the same source, any systematic measurement error probably is not the source of our predictive error. There is no reason to suppose that a randomly chosen subset of the data is more or less prone to measurement error than others. In our judgment, therefore, measurement error is probably not the proximate source of predictive error here.

Temple (1999) documents parameter heterogeneity as a possible problem in parametric econometric analysis. He quotes Arnold Harberger (1987):

> What do Thailand, the Dominican Republic, Zimbabwe, Greece, and Bolivia have in common that merits their being put in the same regression analysis?

Indeed. A point parametric estimate of the marginal effect of a variable on growth papers over true variability, since, such marginal effects may, for example, vary by country.[6] Further, whether such a variable has a statistically significant effect on growth may itself be subject to the choice of sample. Thus, using statistical significance to identify robust correlates of growth may be erroneous. EBA is a

---

[6]We will illustrate this point in the next chapter.

fundamentally parametric approach using statistical significance to identify robust correlates of growth. The variables we choose as robust predictor candidates are themselves the outcome of EBA analysis. This choice itself may therefore be prone to error. Not all variables may be significant an all countries while some insignificant variables may be significant in some countries. Therefore, our lack of predictive precision may be attributed to our choice of variables.

Some Machine Learning approaches are, however, perfectly suited to deal with this problem. Recall that decision trees etc. are designed to identify groups of countries for which parameters differ widely. In fact, one of the earlier applications of regression trees in economics was in the growth literature to resolve precisely this problem (Durlauf and Johnson 1995; Durlauf and Quah 1999). Their work suggests that regression analysis may be particularly unsuitable for econometric growth analysis because of parametric uncertainty and indicated the value of algorithmic non-parametric approaches. The recent burgeoning of computing power has eased the technical hurdles for using these algorithmic approaches. We therefore suggest using our cross validated approach to overcome the parameter heterogeneity problem in the empirical economic growth literature. In fact, in the next chapter, we show the validated patterns of heterogeneity in some of the more important predictors of economic growth. Nevertheless, we can probably conclude here that given our methodology, parametric heterogeneity is probably not the most important source of prediction error.

The growth literature acknowledges that empirical growth modeling is rife with specification problems (Temple 1999). Certain variables are statistically significant covariates of economic growth in some specifications but not in others. Leamer (1985) proposed EBA to solve this model uncertainty problem, thus identifying robust correlates of economic growth. Researchers from Levine and Renelt (1992) to Sala-i-Martin (1997) and beyond have taken this advice to heart. Of course, the reader will recall that we justify our choice of variables for predicting economic growth with Sala-i-Martin's (1997) EBA. Nevertheless, the practical gap between the predictions based on these variables and reality remain. This suggests at the very least two problems: that we are not considering some variables or that EBA is not parsimonious enough in including the right variables. We will come back to this latter problem in the next chapter. We, however, do not address the former problem in at least one meaningful way in this book.

Recall from the introduction that we wanted to focus on the predictive power of variables amenable to policy analysis. Our purpose here was to get a sense of how much confidence policy makers can place on the variables that the literature identifies as robust covariates of growth. Our results here suggest—not much. However, our policy lens necessarily ignores some possible long run determinants of growth. After all, policy makers cannot do anything about a country's geography or history. Nevertheless, there is a burgeoning literature on how culture (through institutions) and biology (epigenetics and learning processes) influence economic growth (see Spolaore and Wacziarg 2013 for a nice review of the seminal work in this area). The magnitude of our prediction errors may therefore be partly driven by the lack of geographic and historical variables.[7] Will inclusion of those variables improve predictability? That is an empirical question we plan on taking up in future work, noting here that we at least partially control for the problem by including institutional variables. Institutions are artifacts of history and culture and certainly differ by geography.

Model uncertainty may also be a function of the channels of growth. Inflation and government control of central banks could be two separate proxies. The fact that one or the other may be a salient covariate in some specifications but not in others may reflect the omission of some more direct measure of bad monetary policy. This kind of omitted variable bias may not be overcome by EBA approaches (Pritchett 1998). Omitted variable and simultaneity biases plague the empirical growth literature because researchers tend to use parametric approaches.

The growth literature is also particularly vulnerable to endogeneity problems. These can arise in two ways. For example, schooling may influence growth, but growth also enables schooling. Lagging or first differencing variables may resolve this problem. Indeed, lagging variables and averaging variables over 5 year periods (we do this as well) is quite standard in the literature. However, it does not address a second, more serious problem. To illustrate—Democracy may affect both economic growth and schooling so that if it is omitted, point parameter estimates of the marginal effect of schooling on growth (even after lagging schooling) would be both biased and inconsistent. Seminal work by Barro and Sala-i-Martin (1995) use instrumental variable approaches to avoid these problems. However, econometric techniques that attempt to solve the problem of endogeneity by using instrumental variables are problematic because the consistency of the estimated marginal effects depends critically on the strength and validity of the instruments. Given the complexity of growth, however, researchers rapidly run up against a shortage of good instruments (Temple 1999). Even using lagged endogenous variables as instruments (Arellano Bond type modeling) is problematic. The effect of education on growth may take a long time, so lagging education a few periods may not grant exogeneity. A truly exogenous lag may even be outside the dataset. Moreover, serial correlation tests of exogeneity may be insufficient since they are based on estimated rather than true residuals. In fact, some of these instruments may actually lead to biased estimates (Bazzi and Clemens 2013). Further, it is even possible for more sophisticated parametric methods to do *worse* than the traditional ordinary least squares method. In short, endogeneity remains a critical problem in the empirical growth literature. Since Machine Learning algorithms do not require any assumptions about the underlying parametric relations between variables, we cannot point to endogeneity source of prediction error in our models. In fact, to the extent Machine Learning algorithms can ignore underlying parametric assumptions we can also exclude error correlation as a final possible source of prediction error. So why are our models not better at predicting a country's growth rate?

---

[7]Nevertheless, we do include institutional variables since policy makers have some control over historical processes endogenous to institutions.

Leaving historical and geographical variables out of our predictive mix may provide a hint. Might it be that we are not including the right variables as predictors? That is certainly plausible. Keep in mind though that we have chosen Sala-i-Martin's (1997) robust covariates of economic growth as our predictors. He did not choose those variables out of a hat. Rather the choice was driven by underlying theoretical models. Thus, if we are not including the right variables then quite possibly we do not have a *complete* growth theory. In the next chapter, we intensify our search for such variables within more recent developments in the growth literature.

It may be equally possible that there is no *unique* theory of growth that can predict growth in all countries. Recall, though, the tree based algorithms perform better than the others. Tree based algorithms (particularly the Random Forest) are particularly useful in determining out-of-sample fit if variables affected growth in different countries differently; i.e. when there may be no unique growth model for all countries. Thus, the tree models should perform well out-of-sample if multiplicity of growth theories were the only problem. Our algorithmic out-of-sample fits leave much to be desired despite this ability to control for parameter heterogeneity. Multiplicity of growth theories may therefore be less of a problem than lack of completeness in current growth theory. In the remaining part of this chapter, we would like to raise a few methodological issues.

Recent guidance from the American Statistical Association suggests that the p-value is misused often. A p-value of 0.01 is usually interpreted as a hard test of a variable's salience in an econometric model. However, all that a p-value of 0.01 suggests is that if a null hypothesis is true, there is a 1% chance of finding another result that is at least as extreme. In short, a significant result may be an outcome of pure chance. Given that academic journals desire to publish significant results, and that there is a low cost of running regressions, one may understand why academics may be disinclined to explore their data further if they happen to chance across a "sexy" significance. EBA can eliminate some of this risk but even that approach presupposes the existence of an underlying 'true' theoretical model. Econometric modeling is therefore particularly susceptible to researcher biases and the incentives of the academic publication market. It is in this area that theoretically agnostic Machine Learning techniques can prove particularly useful.

For example, variables in an econometric model may be significant in terms of p-values. However, such a model may not be validated out-of-sample. Such models are problematic in many ways. A true causal model should be predictive in the sense that it should be generalizable. A model with highly significant variables in a dataset but not in others is an anecdote, not science. Such a modeling approach is particularly susceptible to researcher bias. Machine Learning covariates are validated out-of-sample rather than depend on a researcher's choice judgment. Thus, they can provide some relief from the bias problem as well as instill confidence in the generalizability of a model specification. For example, if for a target A, variables B, C, and D shows better out-of-sample fit than variables B, E, and F then using the former specification is justifiable as a matter of science rather than researcher judgment. Further, this higher out-of-sample fit suggests predictive salience for the specification. After all, what is prediction quality other than being right about something unknown? The unknown is, by definition, out-of-sample![8] This argument can extend to the econometric techniques that do identify causal links by controlling for endogeneity. However, without out-of-sample validation we have no sense of the importance of these causal links *relative* to others. A truly scientific approach would not only identify statistical causal links but also test whether these links work outside the dataset in which the links were determined. Further, just because a variable is significant in an econometric model that controls for endogeneity does not mean that the magnitude of the causal effect is large relative to other causal variables. In other words, the magnitude of the causal effect needs to be evaluated in a relative sense. Some Machine Learning algorithms can do this too. We introduce a methodology for this in the next chapter.

The reader will notice that the only methodology that does any better than the Random Forest is the average of all the algorithms with a MSE of 12.87 in the test sample while reducing the MSE relative to the baseline by 0.127%. This distinction may matter for the practitioner of predictive analytics. Many practitioners look at a low prediction error for a particular algorithm and call it a day. We suggest that prediction errors can be reduced by calculating the average predictions over many different algorithms. This may be especially relevant when predictive capability is particularly important. The reader will however note that the MSE for the average of all the algorithms *here* is not too different from the Random Forest methodology.

It may also come as no surprise to the reader that good fit in the learning sample does not always translate into a good fit in the test sample. For example, note that the single tree has a MSE of 4.8 in the learning sample, which explodes to a 26.02 in the test sample. This points to the danger of "overfitting" and a further advantage of model validation in test samples. Over fitted models give an artificial sense of predictive security—if they were truly good at reducing predictive error they would do so in the test sample as well.

Takeaways

1. The robust correlates of economic growth suggested by EBA (Sala-i-Martin 1997) are not very practical predictors of a country's economic growth. Nevertheless, we learn something about economic growth by eliminating competing theoretical models of growth.
2. The magnitudes of the prediction errors do not seem to be influenced by some of the usual problems (Temple 1999) in the empirical growth literature.
3. A good theoretical model should also be predictive. To the extent that a model based on current growth theory has prediction problems suggests an emphasis on understanding economic growth beyond the usual suspects. Here we suggest looking historical and geographical factors while acknowledging that policy

---

[8]We explicitly focus on prediction in this book by looking at growth 5 years out. But that obviously would not work for cross sectional data. Out of sample tests, however, can measure a specification's predictive power even with cross section data.

makers may not be able to influence the impact of those factors. We will delve deeper into this variable search in the next chapter.

4. Multiplicity of growth theories is less of a problem than lack of completeness in current growth theory.
5. Machine Learning techniques help avoid researcher bias while giving the researcher a sense of the generalizability of a set of results.

References

Barro, R. J., & Sala-I-Martin, X. (1995). *Economic growth.* Boston: McGraw-Hill.

Bazzi, S., & Clemens, M. A. (2013). Blunt instruments: Avoiding common pitfalls in identifying the causes of economic growth. *American Economic Journal: Macroeconomics, 5*(2), 152–186.

Durlauf, S. N., & Johnson, P. A. (1995). Multiple regimes and cross country growth behavior. *Journal of Applied Econometrics, 10*(4), 365–384.

Durlauf, S. N., & Quah, D. T. (1999). The new empirics of economic growth. In J. T. Taylor, W. Michael, J. T. Taylor, & M. Woodford (Eds.), *Handbook of macroeconomics* (Vol. 1A, pp. 235–308). Philadelphia: Elsevier.

Harberger, A. C. (1987). Comment. In S. Fischer (Ed.), *NBER macroeconomics annual* (pp. 255–258). London: MIT Press.

Leamer, E. E. (1985). Sensitivity analysis would help. *American Economic Review, 73*(1), 308–313.

Levine, R., & Renelt, D. (1992). A sensitivity analysis of cross country growth analysis. *American Economic Review, 82*(4), 942–963.

Pritchett, L. (1998). *Patterns of economic growth: Hills, plateaus, mountains, and plains,* Manuscript, World Bank.

Sala-i-Martin, X. (1997). I just ran four million regressions. *American Economic Review, 87,* 178–183.

Spolaore, E., & Wacziarg, R. (2013). How deep are the roots of economic development? *Journal of Economic Literature, 51*(2), 325–369. https://www.aeaweb.org/articles?id=10.1257/jel.51.2.325.

Summers, R., & Heston, A. (1988). A new set of international comparisons of real product and price level estimates for 130 countries, 1950–1985. *Review of Income Wealth, 35*(1), 1–25.

Temple, J. (1999, March). The new growth evidence. *Journal of Economic Literature, 37,* 112–156.

Chapter 5
Predicting Economic Growth: Which Variables Matter

In this chapter, we present our findings and get a better sense of the "best" predictors of economic growth. Some Machine Learning techniques allow us to rank variables according to their predictive salience. We suggest that the search for causality should begin among these more predictively salient covariates. More importantly though, these techniques allow us to eliminate "predictors" that are *unlikely* to be causal.

The variables we find to be important in predicting growth appear to not be in line with the institutional literature but support a more neo-classical understanding of economic growth. We suggest that these findings really hint at potential channels of economic development and growth. We reinforce this idea in the next chapter where we look at predicting recessions.

We report partial dependence plots (PDPs) for variables of interest. The PDPs give a sense of the threshold effects of a discrete change in a variable on economic growth. In addition, by iterating over the entire distribution, the PDPs allow the effects to be more complicated than suggested by simple linear models. A simple parametric point estimate necessarily glosses over the possibility that a covariate's marginal effect on economic growth may be heterogeneous over the range of values for that covariate. This capability directly addresses Temple's (1999) parameter heterogeneity lament referred to in the previous chapter. The outer tick marks in the PDPs mark fixed increments of the value of each variable, while the inner tick marks indicate the cutoffs for each decile of the distribution of the variable. The vertical axis measures the growth rate. These plots help visualize how a change in a

© The Author(s) 2017
A. Basuchoudhary et al., *Machine-learning Techniques in Economics*,
SpringerBriefs in Economics, https://doi.org/10.1007/978-3-319-69014-8_5

covariate affects economic growth in the learning sample.[1] We report the plots over three different algorithms, Bagging, Boosting, and Random Forests. We can discern some of the more obvious policy implications with PDPs.

The basic methodology for ranking variables in order of how well they predict is quite simple. For each type of algorithm, the program removes a variable and reports how much the MSE then increases in the learning sample model. Variables whose removal increases MSE the most are the most important predictors. This allows the algorithm to rank order variables according to their prediction importance.

Before reporting and discussing our results, we need to make a couple of points about applying the variable importance ranking methodology in this book. First, the innate methodology of the ANN makes variable ranking impossible.[2] Second, we report the average increase in Model MSE rather than the individual increases in order to capture the information from all the algorithms. We calculate and report the percentage increase in model-MSE if a variable is removed from the model for each kind of algorithm for which variable ranking is possible. Then we average those MSEs and rank the variables according to these numbers. For example, if LagInvestment/GDP is removed from the model, the MSE increases by 7.21% in the Tree method, by 2.33% in the Forest method, by 1.13% in the Bagging method, and by 43.41% in the Boosting method. The average of these is numbers, 13.21% is in the last column of Table 5.1. The variables are the ranked according to this average measure of importance in Table 5.1. We turn to this table now.

The discussion below proceeds along the following lines: we evaluate variables that are traditionally considered to be salient in the empirical and theoretical growth literature according to their predictive salience. We then look at predictively salient correlates of economic growth that can be influenced by policy.

---

[1]Partial dependence plots display the marginal effect of variable x_k conditional on the observed values of all of the other variables, $(x_{1-k}, x_{2-k}, \ldots x_{n-k})$ Specifically, it plots the graph of the function:

$$\hat{f}(x) = \frac{1}{n} \sum_{i=1}^{n} f(x_k, x_{i,-k}),$$

where the summand, $f(x_k, x_{i,-k})$, is defined by:

$$f(x) = \ln[p_1(x)] - \frac{1}{C} \sum_{s=1}^{C} \ln[p_s(x)].$$

Here, s indexes the classes of the target variable and C is the total number of classes. So, if an increase in a given factor, x_k, increases the probability of state failure ($s = 1$), then the value of the function plotted by the PDP also increases.

[2]We should recall from Table 4.1 in the previous chapter that the ANN algorithms performed particularly poorly both in-sample and out-of-sample. A better predictive showing would have indicated that the covariates of growth operate in interactive ways, but it would have been impossible to use the variable rankings as a source of analytical information. Moreover, the relative failure of the ANN algorithms suggest that the covariates of growth used here have more direct channels of influence.

Table 5.1 Ranking growth predictors

Variable	Tree	Forest	Bagging	Boosting	Average
LagInvestment/GDP	7.213814	2.330242	1.132886	43.41862	13.52389
LagGDPpcGrowth	6.391531	2.605418	2.799576	31.18467	10.7453
LagTermsOfTrade	10.83415	2.758283	2.725816	8.594572	6.228205
LagFDI_In_GDP	6.565904	2.503886	1.509351	6.167281	4.186605
LagLifeExpectancy	3.608791	6.266309	2.977885	0.785733	3.409679
LagImportsGDP	5.191559	3.693854	0.918352	2.122532	2.981574
LagPhonesPC	3.222595	4.825022	2.002104	0.464815	2.628634
LagDependency	3.30342	5.22147	1.389975	0.217326	2.533048
LagDemocracy	2.595834	3.358017	1.615129	2.551995	2.530244
LagPopulationGrowth	2.971569	2.522636	1.200022	3.16817	2.465599
LagGDPpc	3.782944	4.075347	1.522695	0	2.345247
LagSecEnrollRatePCT	1.932597	4.660113	1.505975	0	2.024671
LagTradeGDP	4.389639	3.01447	0.431308	0.048304	1.97093
LagLaborForceParticipation	2.27813	3.827211	0.639334	0.482488	1.806791
LagRuralPopulationPCT	2.664254	3.094718	0.864459	0	1.655858
LagExportPrices	2.692738	2.506973	1.184889	0	1.59615
LagPopulation	0.916803	3.945999	0.968242	0	1.457761
LagFemaleLaborForce	1.71439	3.113742	0.466716	0.108048	1.350724
LagImportPrices	2.186627	2.423107	0.777558	0	1.346823
LagMoneyGDP	1.669132	2.521166	0.619094	0.158933	1.242081
LagGini	2.075124	2.086685	0.71202	0	1.218457
LagAidAssistGDP	2.039484	1.909016	0.511311	0.242548	1.17559
LagExportsGDP	1.397084	2.520725	0.412994	0.023152	1.088489
LagCredibility	1.844861	1.587284	0.352969	0.022353	0.951867
LagTransparency	1.324525	2.007088	0.434475	0	0.941522
LagConsumptionGDP	1.271571	1.677932	0.498525	0.03296	0.870247
LagLendingInterestRate	1.847777	1.214629	0.235496	0	0.824475
LagFDI_Out_GDP	1.081785	1.927486	0.257928	0	0.8168
LagPrimCommodExports	0.718586	1.745085	0.448696	0.142749	0.763779
LagIndustryGDP	0.797948	1.733039	0.518212	0	0.7623
LagInflationCPI	0.839938	1.598046	0.454356	0.029781	0.73053
LagWithinInstab	0.867909	1.468009	0.364357	0	0.675069
LagMilitaryGDP	0.99299	1.264055	0.198774	0	0.613955
LagInterestSpread	0.802032	1.073479	0.230268	0.032965	0.534686
LagSchoolExpendGDP	1.185748	0.803548	0.138596	0	0.531973
LagMoneyGrowth	0.830225	0.887818	0.347397	0	0.51636
LagSystem	0.329853	1.594888	0.099513	0	0.506063
LagTerror	0.810216	0.976091	0.233624	0	0.504983
LagNetGovernmentGDP	0.791167	0.843733	0.135982	0	0.442721
LagProtest	0.564659	0.398138	0.11633	0	0.269782
LagExecutive	0.54805	0.478747	0.044381	0	0.267795
LagRealInterestRate	0.35276	0.45008	0.130056	0	0.233224

(continued)

Table 5.1 (continued)

Variable	Tree	Forest	Bagging	Boosting	Average
LagRegime	0.274454	0.188378	0.103687	0	0.14163
LagRegimeInstab	0.059047	0.164779	0.031004	0	0.063708
LagNonethnicConflict	0.138747	0.044291	0.007701	0	0.047685
LagEthnicConflict	0.087039	0.088969	0.007565	0	0.045893

5.1 Evaluating Traditional Variables

Predictive models are not causal. However, predictability should be a consequence of a good causal model. For example, we know that the effect of mass on gravity is a good causal model because the model correctly *predicts* how fast an object moves because of gravitational forces. Prediction is a test of generalizability. Consequently, variables that are poor predictors of a phenomenon are also unlikely causal candidates. Let us look at those variables that, in the empirical literature, are supposed to be robust covariates of growth. Are they all good predictors of growth?

A quick perusal of Table 5.1 reveals that Per capita GDP does not make it into the top ten predictors. The causal effect of per capita GDP on growth is questionable. Does this mean that past GDP levels are irrelevant for understanding growth? Given the importance of the issue of convergence in the growth literature, the question matters. Of course, the growth literature does not claim that per capita GDP *causes* growth.

Mankiw, Romer and Weil's (1992) seminal work merely notes that initial per capita GDP should have a negative coefficient in a growth regression. This happens because the standard neo-classical growth models suggest that poorer countries should grow faster than richer countries given the same investment levels because the marginal product of capital is higher in poorer countries. In fact, we find this effect in the partial dependence plot for lagged per capita GDP shown in Fig. 5.1. Our predictive model is consistent with the growth regression literature where convergence is about the one thing everyone agrees on even if there is no consensus on the rate of convergence. This should increase the reader's confidence in our results. The confidence will be needed when we turn to the role of institutions in our model!

Spolaore and Wacziarg (2013) survey the literature on historically rooted factors of growth. They suggest that the growth literature has moved away from proximate determinants of growth, factors like physical and human capital accumulation and technology, to more fundamental factors like certain behavioral traits passed on through culture or even genetics. The controversy here seems to be a debate on whether these factors affect productivity directly or indirectly by helping or hindering technology diffusion. Next, we wade into this debate.

The most fundamental models of economic growth emphasize the role of capital accumulation, human and physical capital, and technological diffusion (Romer 2001). Recent investigation of economic growth, however, suggest that initial geographic conditions influence institutions, human capital, and cultural traits, which in turn may affect income and productivity. Moreover, behavioral traits that affect income and productivity may themselves be contextual in the sense that they too are transmitted from one generation to other.

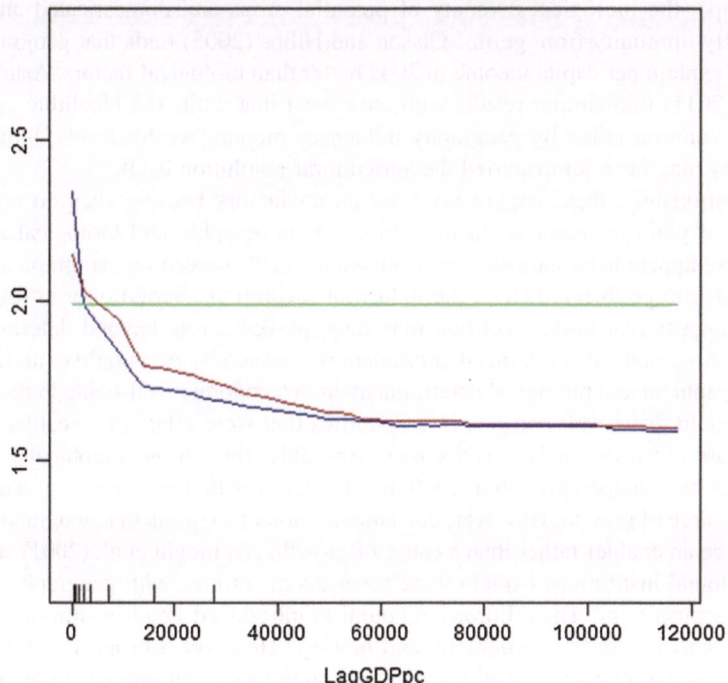

Fig. 5.1 Marginal effect of per capita GDP on economic growth

We have also included variables Sala-i-Martin (1997) suggests are robust correlates of growth. These variables capture many theoretical growth models, including those fundamental models that emphasize the role of capital accumulation, human and physical capital, and technological diffusion. The variables ordained by the fundamental models of growth are expected to directly affect economic growth, while the other variables may affect growth either directly or indirectly. Of the latter, we have deliberately left out variables that capture geography, religion, and colonial history since these cannot be influenced by policy. Given the policy lens of this book, it seemed prudent to include only variables that policy makers can influence. We do, however, include variables that may be *influenced* by geography, history, and culture. We suggested in the previous chapter that the predictive accuracy of machine-learning algorithms may be impaired because of incomplete theoretical models. We will try to get at those "missing" factors through a method of elimination. Let us begin with the effect of geography.

Both Machiavelli (1531) and Montesquieu (1748) noted that geography affects productivity and income. A vast empirical literature has followed since. Sachs (2001) claims that tropical climates inhibit growth because high rainfall depletes soils while pests and parasites affect agricultural and human productivity. Diamond (1997) argues instead that certain geographical regions like Eurasia were more amenable to the Neolithic agricultural revolution because of the nature of the landscape, the biological diversity of potential crops and domesticated animals, and early immunity from germs. Olsson and Hibbs (2005) finds that geographical factors explain per capita income in 2005 better than biological factors. Ashraf and Galor (2011) find similar results with the caveat that while the Neolithic agricultural revolution aided by geography influences modern wealth levels, biological diversity may have jump started the agricultural revolution itself.

Unfortunately, these explanations are unsatisfactory because they do not adequately explain documented history. First, the geographic and biological advantages that appear to explain economic advantage in the preceding paragraph applied to all of Europe. Nevertheless, the industrial revolutions started only in England. This suggests that human volition may have played a role beyond deterministic nature. A second caveat to this determinism was issued by Acemoglu et al. (2001). If geographical and biological determinism drove economic well-being in the 1500, they should drive well-being today. Countries that were relatively wealthy in the 1500s are not necessarily wealthy now. Arguably, though, geographical features that may have helped growth in 1500 may hinder growth now because of a change in the source of growth. However, this latter argument suggests that geography may at best be an enabler rather than a cause of growth. Acemoglu et al. (2001) suggest that colonial institutions explain these reversals of fortune, with geography acting as a co-conspirator. Thus, European colonists introduced good institutions where settling was favored by geography and biology. However, European colonialism resulted in bad extractive institutions where they faced bad biogeographic conditions. Their study therefore places explanatory primacy on institutions over biogeography—part of a long and hallowed literature dating back at least to North and Thomas (1973). However, Engerman and Sokoloff (1997) somewhat preempt the institutional argument by claiming that different factor endowments created differences in the degree of inequality which in turn determined the nature of institutions. More unequal societies had more extractive institutions. This in turn reduced the incentive for investment and innovation relative to societies with a more equal distribution of wealth. Further, Glaeser et al. (2004) suggest that "Europeans who settled in the New World may have brought with them not so much their institutions, but themselves." This latter criticism of the institutional explanation of economic growth suggests that culture may play an important role in explaining and therefore predicting economic growth. How do our results inform this debate?

We do not include biogeographical factors in our data set since we want this book to inform policy decisions and nothing much can be done about biogeography as a matter of policy. The discussion above also suggests that biogeographical determinism may be overblown. Thus, it seems that excluding biogeographical variables as potential predictors is probably not the reason for of our weak predictions. Nevertheless, the literature on the effect of institutions on economic growth, the same literature that suggests that institutions intervene and moderate biogeographical determinism, find little predictive salience in our literature either. We find that other than democracy none of the variables that proxy political or economic institutions have any predictive salience for economic growth.

We find that income inequality as measured by the Gini coefficient is not a very good predictor of economic growth either, ranked 21 in Table 5.1. Neither are Credibility (ranked 24), which captures the direct influence of economic institutions, or Transparency (ranked 25), which captures the influence of political institutions. We find little evidence that supports Engerman and Sokoloff's (1997) argument that income inequality itself drives the direction of institutional evolution away from, or toward, inclusive institutions. If it did then either income inequality, or measures of institutional quality, or both would be predictively salient. In fact, if institutions influenced economic growth directly, measures of institutional quality should have predictive salience. Therefore, to the extent that these kinds of variables show up as consistent correlates of economic growth, their causal influence on economic growth must either be indirect or non-existent. This at least suggests that theoretical and empirical modeling that consider the indirect influence of economic and political variables on economic growth, possibly through cultural transmission mechanisms, should be favored over those that model the effect directly. In fact, Michalopoulos and Papaioannou's (2010) study of African countries suggest that differences in ethnic identity rather than institutional differences may explain economic differences. This is broadly in line with Glaeser et al.'s (2004) comment on the role of European cultural features, rather than institutions, in explaining economic success. Thus, it is to culture we may need to turn to for a more complete growth prediction model.

Anthropological work (see for example Guglielmino et al. 1995) suggest that intergenerationally transmitted behavioral traits show remarkable resilience in the face of shifting contextual sands of institutions, geography, and biology. Moreover, Putterman and Weil (2010) suggest that even if institutions matter for economic growth, they do so because institutional norms are passed down through familial cultural transmission mechanisms. Interestingly Comin, Easterly, and Gong (2010) suggest that historical technological advantages, such as the adoption of some basic technologies as far back as 1000 BC, provide benefits to populations that adopted those technologies earlier rather than when that technology was adopted in the current location of the population. Thus, our prediction errors may well be an artifact of our failure to include variables that capture cultural transmission mechanisms.[3]

We suggest that future work on improving predictive validity should include mechanisms of intergenerational transmission, i.e. culture.[4] To wit, these variables should capture Jablonka and Lamb's (2005) description of four inheritance dimensions: genetic, epigenetic, behavioral, and symbolic. If including these sorts of

---

[3]Basuchoudhary and Cotting (2014) develop a model for understanding cultural transmission mechanisms that show how culture rooted in psychology can shift a society from one equilibrium to another.

[4]Human beings are not solely coolly calculating optimizers or even a collection of consistent behavioral oddities. Those rational actor or behavioral economics models are important. However, people also work in a cultural context. We merely suggest that this cultural context should be part of any understanding of human endeavor. Our algorithms seem to be pointing to the salience of the "dog that did not bark" cultural aspect in the growth literature. We therefore echo Morson and Schapiro's (2017) call for economists to think of human behavior in a cultural context.

variables in a predictive model reduces prediction error, then we would be more confident of both our ability to predict growth (obviously!) and in our theoretical understanding of the process of growth. The reader will note here how Machine Learning algorithms have helped us discern potentially missing components of a comprehensive theory of growth. Thus, even if it turns out that cultural transmission processes do not improve predictive capability, and therefore should not be part of a comprehensive causal theory of growth, the validated Machine Learning *process* can still be a tool in the search for such a theory.

Before we move on to our section on policy variables in this chapter, we would like to make one last point on the role of intergenerational transmission mechanisms. If familial transmission mechanisms matter for economic success then familial boundaries could hinder learning because these boundaries potentially cut off the learning channel. Spolaore and Wacziarg (2009) find that genetic distance in 1500 (a proxy for genetic distance between indigenous populations) is negatively related to income per capita in 2005. Moreover, genetic distance from technological frontier (the US or Northwestern Europe) also significantly explains economic distance. However, over time the magnitude of this effect declines. Arguably, increased globalization allows people of different familial/ethnic lines to learn from each other, therefore increasing rates of technology diffusion. Increased trade and economic interactions should therefore reduce the magnitude of the negative effect of genetic distance on economic success. We find results broadly consistent with this argument. Notice that incoming FDI as a percentage of GDP (LagFDI_In_GDP) is the fourth most important predictor of economic growth. Incoming Foreign direct investment is a source of foreign technology. Similarly, we find that imports as a percentage of GDP (LagImportsGDP) is the sixth most important predictor of economic growth—imports, too are sources of foreign technology. Moreover, the PDPs in Figs. 5.2 and 5.3 both show that imports and incoming FDI increase economic growth.

Ethnic conflict is a direct measure of ethnic/familial barriers arising out of genetic distance. Presumably, if genetic distance had current salience in explaining differences in economic success then ethnic groups in conflict should predict a fall in economic growth. Out of a group of 46 potential predictors of economic growth, ethnic conflict comes dead last. We suggest that even if genetic distance may have played a role as a barrier to technological diffusion and learning norms in the past it is not so now—in line with Spolaore and Wacziarg (2009) finding of the diminishing magnitude of the genetic distance effect. Nevertheless, we suggest caution in thinking of this phenomenon as permanent given the apparent rising tide of ethnonationalistic and antiglobalization forces.

We have argued that the cultural transmission mechanism may be a missing link in a comprehensive causal theory of economic growth. If true, then as and when we understand this process better, economists may ally with psychologists, sociologists, and anthropologists to develop robust and inclusive economic development policies. In the meantime, validated Machine Learning can continue to provide insight into which policy tools are more effective than others in encouraging economic growth based on what we do know about economic growth. We turn to these policy tools next.

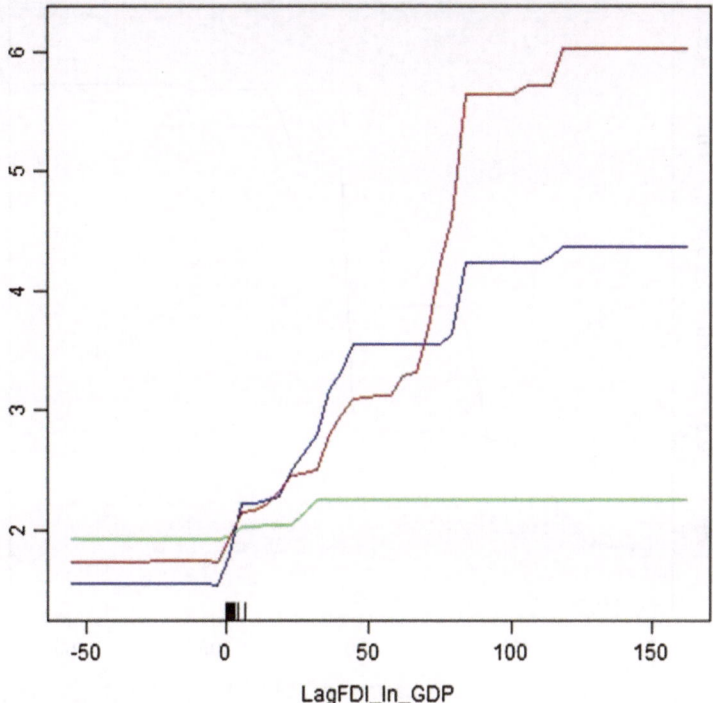

Fig. 5.2 Marginal effect of incoming FDI on economic growth

5.2   Policy Levers

We have included all the variables suggested by Sala-i-Martin's (1997) EBA analysis in our Machine Learning algorithm, with the notable exceptions of Colonial History and Religion. Our argument for not including those sorts of variables is that there are no ethical ways of changing people's religion or any realistic way of wiping out history. Nevertheless, we have included variables that channel the potential effects of religion or history. Presumably, these effects are embodied in the current socio-institutional structure of countries. The sort of social and institutional variables that capture the current socio-institutional structure of countries as recommended by Sala-i-Martin's (1997) EBA results *are* included in our data set. Which of these variables are the best policy levers for sustained economic growth?

This question is part of an ongoing debate generated by the decline of the Washington Consensus; i.e. the "stabilize, privatize, and liberalize" mantra. The World Bank these days suggests the need for humility, policy diversity, selective modest reforms, and experimentation (Rodrik 2006). Rodrik further notes that growth policy should focus on some of the "binding constraints on economic growth rather than take a laundry list approach a la Washington Consensus." He

Partial Dependence on LagImportsGDP

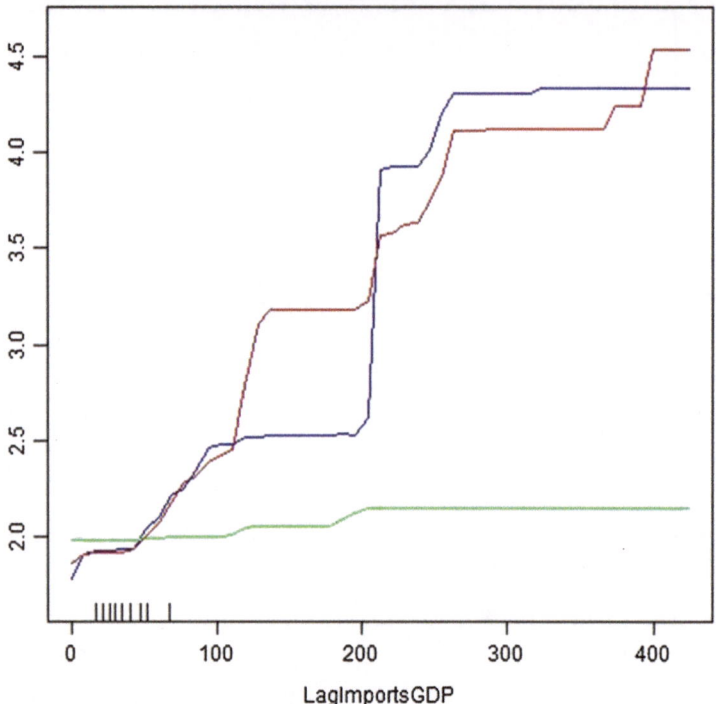

LagImportsGDP

Fig. 5.3 Marginal effect of imports on economic growth

Table 5.2 Top ten predictors of economic growth in our model	Rank	Variable
	1	LagInvestment/GDP
	2	LagGDPpcGrowth
	3	LagTermsOfTrade
	4	LagFDI_In_GDP
	5	LagLifeExpectancy
	6	LagImportsGDP
	7	LagPhonesPC
	8	LagDependency
	9	LagDemocracy
	10	LagPopulationGrowth

then suggests that common sense and economic analysis can help guide identify and unbind these constraints. We attempt at just such an analysis to identify these binding constraints.

Table 5.2 highlights the top ten predictors of economic growth in terms of how much each contributes to increasing out-of-sample fit. This suggests that of the usual covariates of economic growth these affect growth more than others.

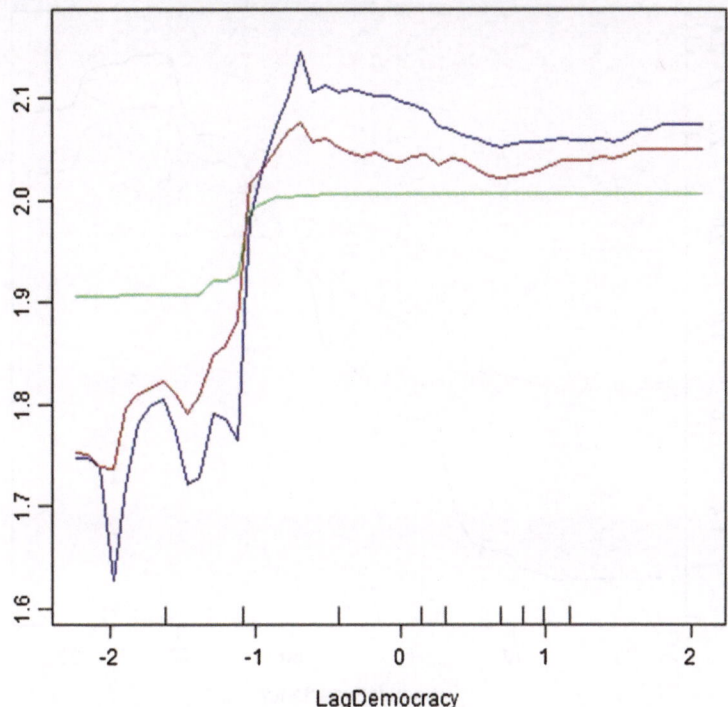

Fig. 5.4 Marginal effect of democracy on economic growth

Therefore, if these variables can be manipulated then they are more likely to affect growth than the others. We choose the top ten merely as a proof of concept. A policy maker may choose more of less depending on which variables are more susceptible to policy manipulations.

Notice that Democracy is the only institutional variable of some predictive salience (ranked 9 in Table 5.2). This is therefore an institution that can be influenced as an economic growth policy. A standard parametric point estimate would suggest a direct relationship between democracy and economic growth. Variable transformations such as squaring the democracy variable even give us information about the nature of the relationship. The PDPs, though, tell us how these relationships change over the range of the predictor variable. Thus, notice in Fig. 5.4, the effect of Democracy on economic growth is highest for relatively small improvements in democracy. This suggests that even a small amount of political liberalization can have large effects on economic growth. Further increases in Democracy has little effect on economic growth, suggesting democracy results in an equilibrium *shift* from low to high growth.

Let us first focus on four of these top ten variables because together they suggest the predictive salience of endogenous growth models that highlight the importance of

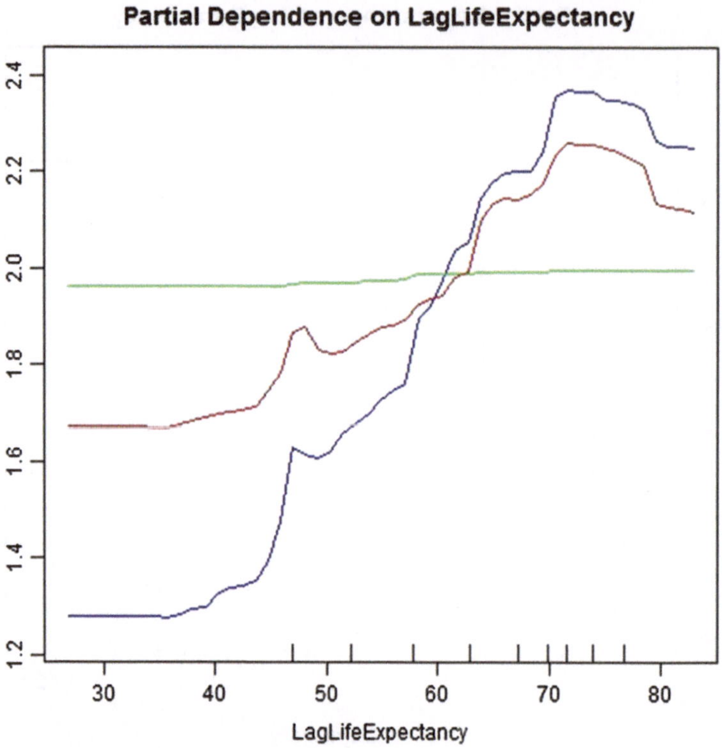

Partial Dependence on LagLifeExpectancy

Fig. 5.5 Marginal effect of life expectancy on economic growth

physical capital, human capital, technology diffusion and labor.[5] These are Investment as a proportion of GDP (LagInvestment/GDP), Life expectancy (LagLifeExpectancy), Phones per capita (LagPhonesPC) and the population growth rate (LagPopulationGrowth). Investment as a proportion of GDP is the most important factor predicting economic growth, and is the usual measure of physical capital. Life expectancy, which is a measure of human capital, is also an important predictor of economic growth, ranked 5 in Table 5.2, as is Phones per capita, ranked 7, which is a proxy for ease of communication. Finally, population growth rate, ranked 9, is also important for growth. A healthy, educated labor force combined with physical capital in countries where technological diffusion is made easier by communication technology is a good way for countries to start on the path of economic growth.[6] Next, let us look at these variables separately, and see what we can learn from their individual PDPS.

The PDP for Life Expectancy in Fig. 5.5 suggests a secularly increasing relationship between life expectancy and growth. Thus, as a matter of policy,

---

[5]Please see Chap. 3 in Romer's (2001) Advanced Macroeconomics textbook for a detailed yet simple exposition of the basic modeling structure.

[6]After all, per capita secondary school enrollment is the 12th most important predictor of economic growth.

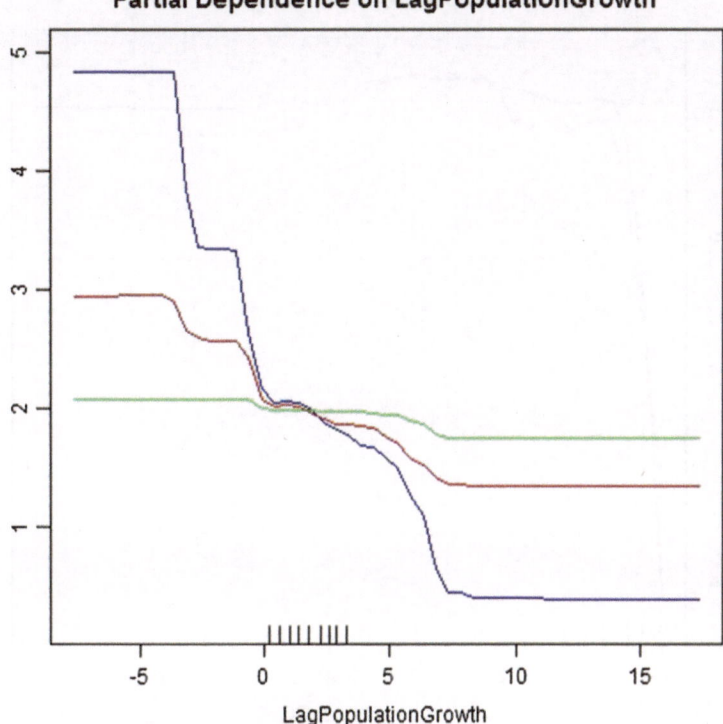

Partial Dependence on LagPopulationGrowth

LagPopulationGrowth

Fig. 5.6 Marginal effect of population growth on economic growth

investments in health care that increase life expectancy of a population appear to have a large effect on economic growth.

A growing population, controlling for other variables, on the other hand, predicts an equilibrium shift from high to low economic growth. This somewhat Malthusian finding should however be tempered with the knowledge that population growth tends to be endogenous to economic growth—higher economic growth tends to reduce population growth rates. The finding here (Fig. 5.6), coupled with Fig. 5.5, suggests that while a growing population may reduce economic growth, a healthy population increases economic growth rates. The policy emphasis here is on the quality rather than the quantity of people in a country. Thus, economic success for populous countries like India may be dependent on investments in health care.

Improvements in communication can help technology diffusion. This effect is captured by Fig. 5.7, which shows the marginal effect of Phones per capita on economic growth. We notice that small investments in communication may increase economic growth by a lot, but this effect levels off over larger investments, suggesting an equilibrium shift from low to high growth investments in communications. Moreover, we seem to get a more complete picture when we combine our view of Fig. 5.6 with Figs. 5.2 and 5.3. Figures 5.2 and 5.3 suggest that trade deficits

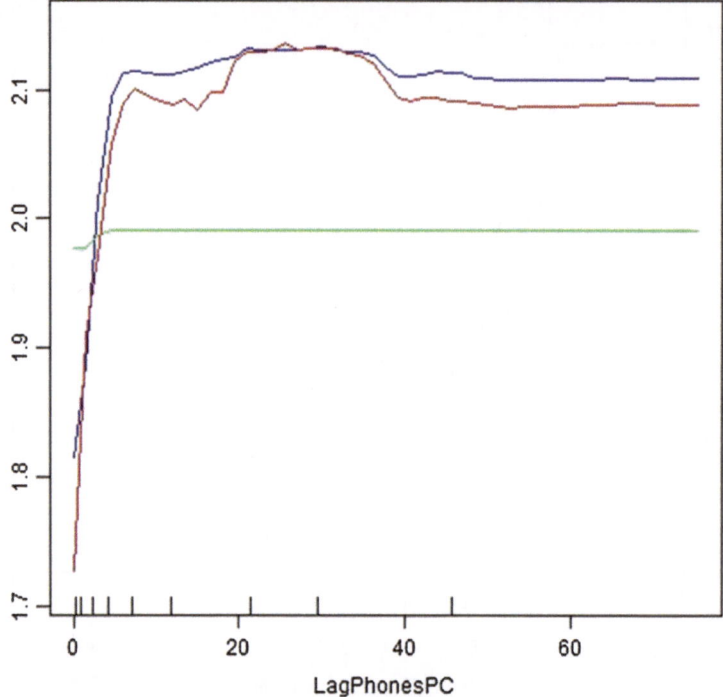

Fig. 5.7 Marginal effect of phones per capita on economic growth

financed by foreign direct investment improve economic growth, as does imports. Of course, foreign direct investment in a country and imports are also a source of technology. Thus, broadly, these effects may be working to improve growth by increasing the likelihood of technology diffusion and endogenous growth. Moreover, Figs. 5.2 and 5.3 suggest that while trade deficits balanced by foreign FDI may in fact enhance growth over a certain range of values, this effect levels off. This in turn suggests globalization leads to an equilibrium shift from low to high growth. These equilibrium shifts may also be interpreted as increasing returns from initial changes in these variables followed by stagnant returns. To the extent stagnant returns do not seem to present themselves for investments in health care, it may suggest secularly increasing investment in health care rather than say communication technology as a matter of policy. We also find evidence of an equilibrium shift from low to high growth because of increasing investments (representing physical assets) as a percentage of GDP. This PDP is shown in Fig. 5.8.

Dependency ratios are an anomaly in our findings (Fig. 5.9). The dependency ratio does not show a clear equilibrium shift effect on economic growth. Rather the PDP for dependency ratios suggests that initially increasing dependency ratios increase economic growth and then decrease it. That a higher dependency ratio might decrease growth by diverting resources from production to maintaining an

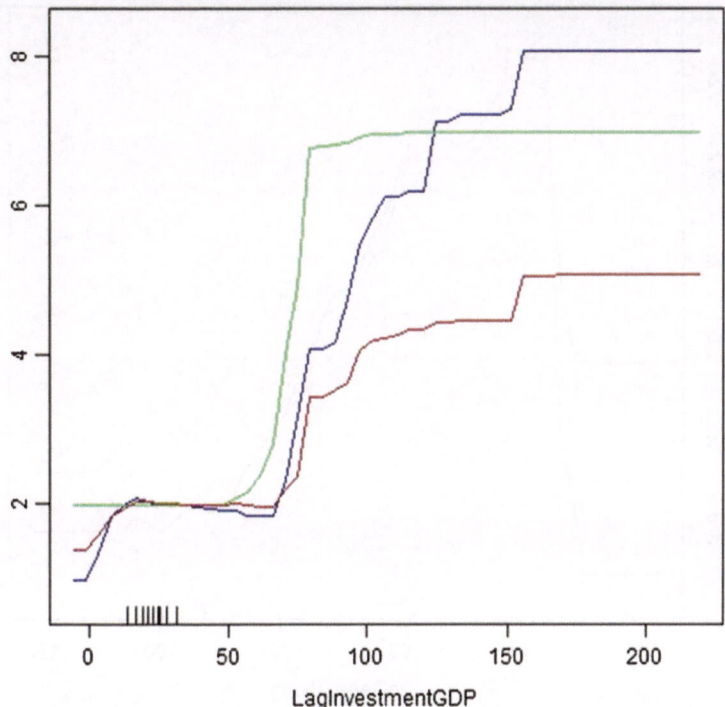

Fig. 5.8 Marginal effect of investment as a percentage of GDP on economic growth

unproductive population seems plausible. However, if that is true then it is unclear how an increase in dependency ratios can lead to higher growth in any range of dependency ratio values. This finding therefore does not fit in any neat category, either intrinsically as we noted above, or extrinsically in the sense of where it fits among all the theoretical and empirical models of economic growth. Nevertheless, it is an important predictor of economic growth and should find a place in a unified theory of economic growth.

Additionally, as we see in Fig. 5.10, worsening terms of trade reduces growth. Little may be done about this in a globalized economy short of instituting policies that reduce the effect of terms of trade shocks without e.g. affecting incoming FDI. This, somewhat in opposition to received wisdom, suggests current account controls and capital account liberalization.

The PDPs in Figs. 5.2, 5.3, 5.4, 5.6, 5.7, 5.8, and 5.10 do have a certain thematic unity. All of them suggest that the growth process shows an equilibrium shift from low to high growth because of changes in investment, human capital, and elements that help technological diffusion. Life expectancy (Fig. 5.5) is the only exception in the sense that economic growth seems to increase secularly with life expectancy. Recall further that investment, technological diffusion, and human capital are all

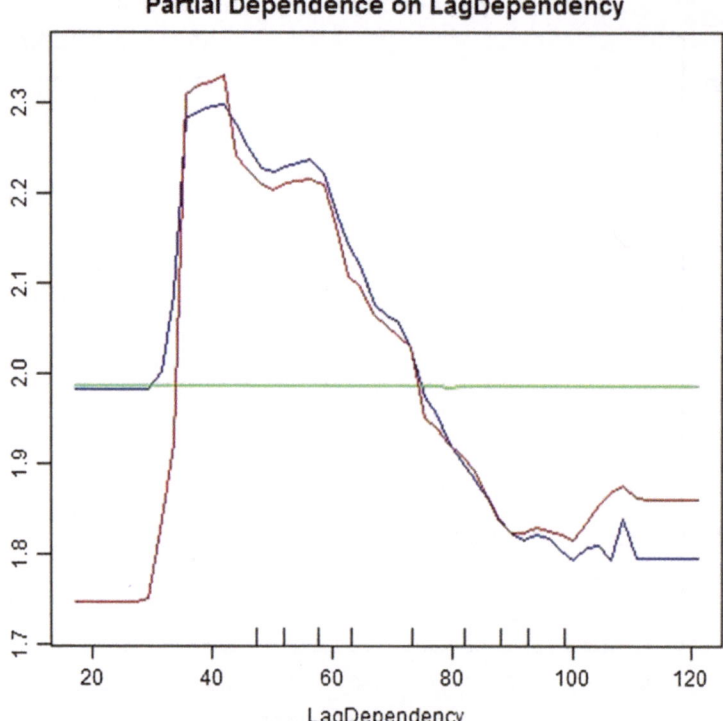

Fig. 5.9 Marginal effect of dependency on economic growth

features of endogenous growth models. The variables representing these ideas are in the top ten predictors of economic growth. Moreover, we note from Table 5.1 that past growth is the second most important predictor of economic growth. Endogenous growth, once initiated is self-perpetuating. Figure 5.11 suggests such an equilibrium shift; low growth countries show persistent low growth while if a country once shifted into high growth continues to sustain growth. Conversely, it is possible for countries to shift back from high to low growth and stay there. Thus, overall, we seem to find that endogenous growth models are the most predictive of the theoretical approaches to growth models.

We therefore suggest the search for the binding constraints on growth, be focused on endogenous growth models. Further, if institutions or income inequality matter (and they do not seem to) then, with the exception of democracy, they probably matter by influencing endogenous growth. Rodrik (2006) agrees by noting that the cross-national literature does not suggest any causal link between institutions and economic growth. Nevertheless, institutions are both formed by, and inform, culture. Our results here therefore suggest that the absence of a better understanding of cultural transmission mechanisms probably creates a significant blind spot for a policy maker. At the same time, this finding highlights the possibility that

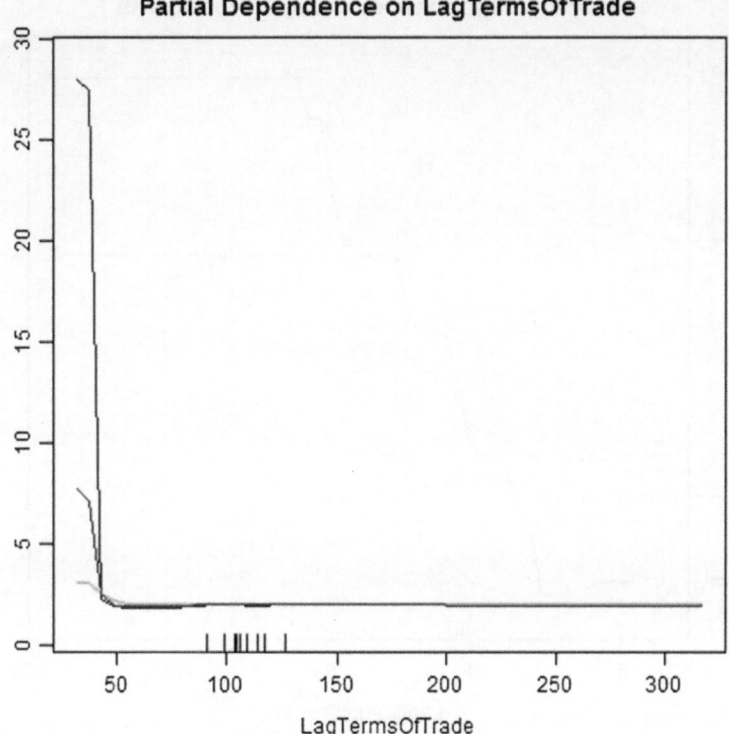

Fig. 5.10   Marginal effect of terms of trade on economic growth

differences in *policy* approaches to unshackling the binding constraints on economic growth probably ought to be rooted in an understanding of culture.

Finally, Sachs et al. (2004) suggest foreign aid as a way to escape a poverty trap. We do find evidence of such poverty traps among the PDPs. However, Sachs et al.'s response that foreign aid can solve this problem needs to be tempered by our finding that foreign aid is not a very good predictor of economic growth. This may be because foreign aid, as it is currently structured, does not actually relax the binding constraints that we have identified.

Coda

We should note that ranking important variables can be a source of positive policy. Thus, a policy maker can use the results reported here to make a case for expanding health care and private investment over, say, foreign aid. However, the policy maker may be unconvinced about this policy because of the overall poor predictive power of these algorithms. We would respond by saying that at least now the policy maker knows how much confidence they should have in the efficacy of policy based on our current knowledge of economic growth! In short, we would have a sense of deep humility about policy implications and our understanding of economic growth. We can however be more confident about negative policy, that is, we

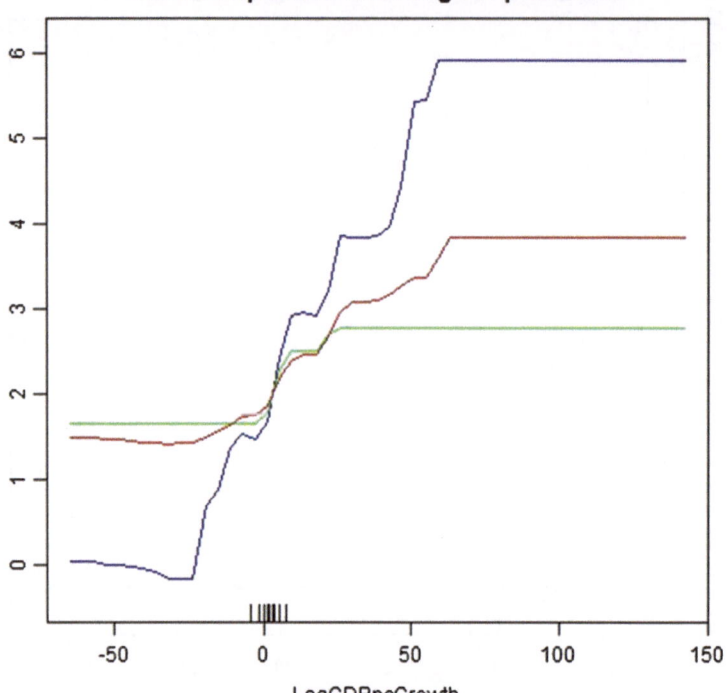

Fig. 5.11 Marginal effect of past growth on economic growth

know which variables or policy levers do *not* work. Thus, clearly foreign aid, as it is currently structured, does not appear to have any consequential impact on economic growth. Further study may suggest restructuring foreign aid toward helping, say, incoming private investment, or even eliminating it as a source of economic development. Nevertheless, knowledge of what does not contribute to economic growth is important too. Machine learning can therefore serve as a methodological framework for evaluating the relative salience of theory and policy through a predictive lens.

Takeaways

1. It is important for future research to focus on cultural and other intergenerational information transfer mechanisms to get a more complete understanding of growth since geography, history, and institutions (other than democracy) appear to have little predictive value in terms of out-of-sample fit.
2. Many of the variables that EBA suggests as robust correlates of economic growth are not good predictors. This suggests that EBA is not parsimonious enough in helping focus on potential causal factors of economic growth.

3. The variables that do predict growth well suggest that the process of growth is not a smooth secular increase—rather each country faces an equilibrium shift from low to high growth. Thus, poverty traps are a very real possibility.
4. Trade deficits financed by FDI increases growth as do investments in health care and communication. Moreover, policy makers should be cautious about potential terms of trade shocks. Particularly for developing countries, these results suggest capital account liberalization with restraints in the current account.
5. Democracy is the only predictive institutional variable.
6. In the absence of variables that capture intergenerational cultural transmission mechanisms, endogenous growth models appear to explain the binding constraints on growth common to all countries. This suggests that policy should focus on easing these constraints rather than relying on a policy maker's gut instinct.
7. Machine learning algorithms can be useful for *eliminating* policy levers that do not work.

References

Acemoglu, D., Johnson, S., & Robinson, J. A. (2001). The colonial origins of comparative development: An empirical investigation. *American Economic Review, 91*(5), 1369–1401.

Ashraf, Q., & Galor, O. (2011). Dynamics and stagnation in the MAlthusian epoch. *American Economic Review, 101*(5), 2003–2041.

Basuchoudhary, A., & Cotting, D. (2014). Cultural assimilation: The political economy of psychology as an evolutionary game theoretic dynamic. *Evolutionary Behavioral Sciences, 8* (3), 209–222.

Comin, D., Easterly, W., & Gong, E. (2010). Was the wealth of nations determined in 1000 BC? *American Economic Journal: Macroeconomics, 2*(3), 65–97.

Diamond, J. (1997). *Guns, germs, and steel: The fates of human societies.* New York: Norton.

Engerman, S. L., & Sokoloff, K. L. (1997). Factor endowments, institutions, and differential paths of growth among new world economies: A view from economic Historians of the United States. In S. Harber (Ed.), *How Latin America fell behind* (pp. 260–304). Stanford: Stanford University Press.

Glaeser, E. L., La Porta, R., Lopez-di-Silanes, F., & Shleifer, A. (2004). Do institutions cause growth? *Journal of Economic Growth, 9*(3), 271–303.

Guglielmino, C. R., Viganotti, C., Hewlett, B., & Cavalli-Sforza, L. L. (1995). Cultural variation in Africa: Role of mechanisms of transmission and adaptation. *Proceedings of the national academy of sciences of the United States of America, 92*(16), 7585–7589.

Jablonka, E., & Lamb, M. J. (2005). *Evolution in four dimensions: Genetic, epigenetic, behavioral, and symbolic variatin in the history of life.* Cambridge: MIT Press.

Machiavelli, N. (1531/2009). *Discourses on Livy.* Oxford: Oxford University Press.

Mankiw, G. N., Romer, D., & David, N. W. (1992). A contribution to the empirics of economic growth. *Quarterly Journal of Economics, 107*(2), 407–437.

Michalopoulos, S., & Papaioannou, E. (2010). *Divide and rule or the rule of the divided? Evidence from Africa.* Center For Economic and Policy Research Discussion Paper 8088.

Montesquieu, C. (1748/1989). *The spirit of the laws.* Cambridge: Cambridge University Press.

Morson, G. S., & Schapiro, M. (2017). *Cents and sensibility.* Princeton: Princeton University Press.

North, D. C., & Thomas, R. P. (1973). *The rise of the western world: A new economic history.* Cambridge: Cambridge University Press.

Olsson, O., & Hibbs, D. A. (2005). Biogeography and long run economic development. *European Economic Review, 49*(4), 909–938.

Putterman, L., & Weil, D. N. (2010). Post-1500 population flows and the long run determinants of economic growth and inequality. *Quarterly Journal of Economics, 125*(4), 1627–1682.

Rodrik, D. (2006). Goodbye Washington Consensus, Hello Washington Confusion? A review of the world banks "Economic growth in the 1990's: Learning From a Decade of Reform". *Journal of Economic Literature, XLIV*, 973–987.

Romer, D. (2001). *Advanced macroeconomics*. New York: McGraw Hill Education.

Sachs, J. D. (2001). *Tropical underdevelopment*. NBER Research Working Paper 8119.

Sachs, J. D., McArthur, J. W., Schmidt-Traub, G., Kruk, M., Bahadur, C., Faye, M., & McCord, G. (2004). Ending Africa's poverty trap. *Brookings Papers on Economic Activity, 1*, 117–216.

Sala-i-Martin, X. (1997). I just ran four million regressions. *American Economic Review, 87*, 178–183.

Spolaore, E., & Wacziarg, R. (2009). The diffusion of development. *Quarterly Journal of Economics, 124*(2), 469–529.

Spolaore, E., & Wacziarg, R. (2013). How deep are the roots of economic development? *Journal of Economic Literature, 51*(2), 325–369. https://www.aeaweb.org/articles?id=10.1257/jel.51.2.325.

Temple, J. (1999, March). The new growth evidence. *Journal of Economic Literature, 37*, 112–156.

Chapter 6
Predicting Recessions: What We Learn from Widening the Goalposts

In the previous chapters, we used Machine Learning to predict economic growth. We learnt that moving toward a more parsimonious and validated list of variables ranked according to their predictive importance can help us understand the process and policy of economic growth a little better. In this chapter, we widen the predictive goal posts to include recessions. A few papers have begun to look at algorithmic methodologies to predict recessions (see for example Berge 2015; Dopke et al. 2017). Their approach is similar to ours in that they identify the most important predictors, and in the case of Dopke, Fritsch, and Pierdzioch (2017), the marginal effects of these indicators on the likelihood of a recession.[1] The papers typically combine leading indicators and allow algorithms to predict recessions. We differ from these works in that we use the covariates of growth, not leading indicators (though there is some overlap), in our algorithms. We do so in order to look at recessions as a *symptom* of problems in the growth process. Moreover, our investigative strategy is more robust in ways we discuss below. What then can we learn about recessions by moving toward a more parsimonious, validated, list of these covariates of economic growth ranked by predictive importance?

In Chap. 4, we noted that we had a large error (as much as 3.6%) in predicting growth. Would our results be more accurate if we widened the predictive goalposts to include two categories of growth—positive and negative? Please note that it is not our intention to compare our accuracy in predicting growth to our accuracy in predicting recessions—growth is a continuous target variable, while recession is a categorical variable. Certainly, theoretically recessions are treated differently from growth. Nevertheless, the two issues are clearly related—recessions being negative growth. Further, differences in the empirical treatment of growth and recessions

---

[1]They, like us choose their model specification by starting with variables arising out of a common understanding of leading indicators typically used to predict recessions. This approach is similar in spirit to ours; we, however, rely on commonly understood covariates of growth since our focus is on understanding the growth process.

© The Author(s) 2017
A. Basuchoudhary et al., *Machine-learning Techniques in Economics*,
SpringerBriefs in Economics, https://doi.org/10.1007/978-3-319-69014-8_6

notwithstanding, the intuitive point that we should be able to predictively hit a wider target better seems plausible. Therefore, after a brief recounting of some stylized facts about macroeconomic fluctuations, we follow the investigative pattern laid out in the previous two chapters. First, we report the results on the most predictive algorithms. Then, we rank the most important predictive variables and present selected partial dependence plots to highlight what we add to the literature on recessions. The basic takeaways from our investigation are at the end of this chapter.

A quick look at US growth rates from the 1960s till 2015 (Fig. 6.1) gives a sense of GDP fluctuations in the United States. Notice that there does not appear to be a discernible pattern in these fluctuations. Between 1965 to 1970, real GDP fell about 3%, from 6.5% to about 3%, but only by 1% (from 4% to 3%) between 1993 and 1995. Growth peaked in 1984 to a little over 7%, but then declined almost continuously until 1991, followed by a steep partial recovery in 1992. The "great moderation" since the 1981–1982 recession did not last forever. In other words, declines in GDP vary both in size and duration, as well as in speed—some declines are faster than others. Therefore, macroeconomists have moved away from trying to discern patterns in business cycles (e.g. Kuznets' 20-year cycles and Kondratiev's 50-year cycles) to thinking of these fluctuations as results of shocks to economic systems. Macroeconomists disagree, however, on the sources of these shocks. We will show how Machine Learning algorithms can contribute to this literature by exalting some of these shocks over others. The only methodological distinction here over the previous chapter is that our predictive target is now a binary choice variable—any negative growth is coded as a recession, 1, and any positive growth is therefore coded as 0. This difference focuses the algorithms approach to minimizing classification error rather than mean square error.

6.1   Predictive Quality

Table 6.1 reports six measures of predictive quality—the receiver operating characteristic (ROC) curve; the area under the receiver operating characteristic curve (AUC); the specificity, or true negative rate; the sensitivity, or true positive rate; the positive predictive value (PPV); the negative predictive value (NPV); and the overall error rate— for seven different techniques—Logit, Single Classification Tree, Bagging, Random Forest, Adaboost, ANN, and an average of overall prediction quality.[2] All these models use imputed values from the Random Forest model in place of missing observations.

For a given predictor, changing the threshold for classifying an observation as 'positive' (a recession) depends on the threshold we choose for the predicted probability of a recession. With a higher threshold, the likelihood that a given incidence of recession is correctly classified goes down (corresponding to a *low sensitivity*, or true

---

[2]Note, for the c-statistic "Average" refers to AUC of the average of the predictions and not the average of the AUCs of the predictions.

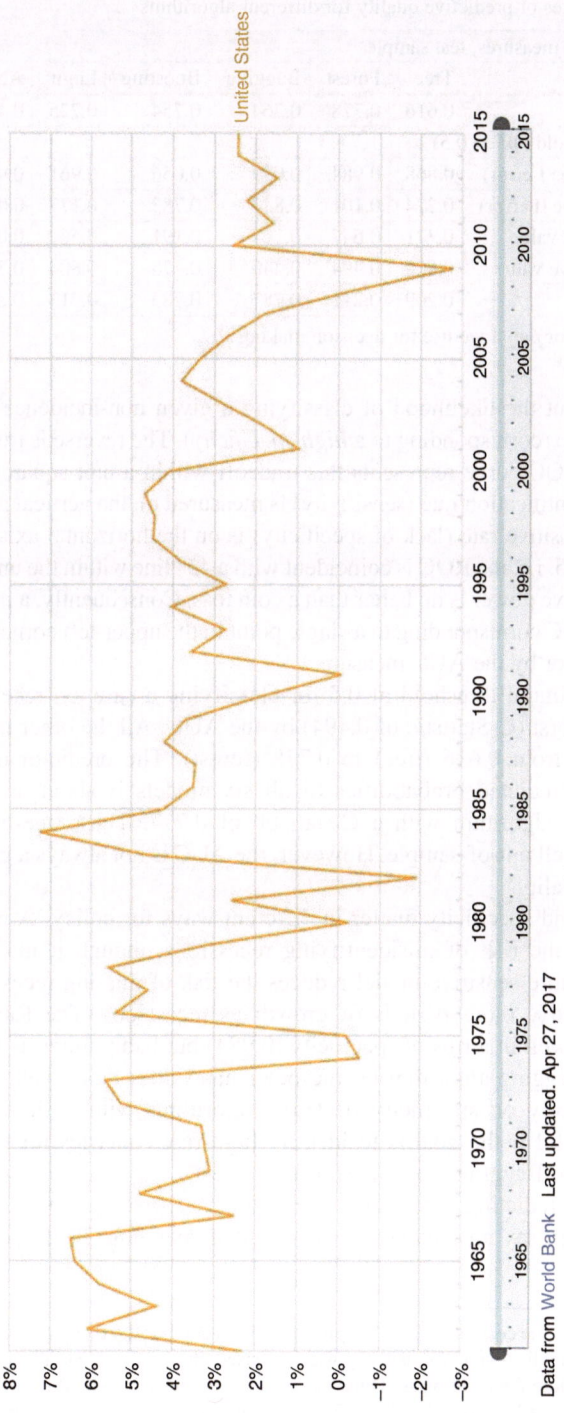

Data from World Bank   Last updated. Apr 27, 2017

Fig. 6.1   US annual growth rates

Table 6.1 Measures of predictive quality for different algorithms

Prediction quality measures, test sample							
Measure	Tree	Forest	Bagging	Boosting	Logit	ANN	Average
C-Statistic	0.616	0.778	0.764	0.754	0.725	0.494	0.777
"Classical" threshold ($p* = 0.5$)							
Specificity (1-Type I error)	0.868	0.984	0.039	0.056	0.961	0.979	0.971
Sensitivity (1-Type II error)	0.274	0.101	0.827	0.783	0.177	0.002	0.159
Positive predicted value	0.371	0.647	0.197	0.191	0.562	0.030	0.611
Negative predictive value	0.808	0.794	0.440	0.478	0.804	0.775	0.802
Overall error rate	0.209	0.211	0.787	0.783	0.213	0.237	0.209

*Threshold value for p that we use for decision making

positive rate), but the likelihood of classifying a given non-incidence of recession correctly goes up (corresponding to a **high** *specificity*). The reverse is true for a lower threshold. An ROC curve represents this tradeoff within a unit square.[3] A model's true positive identification rate (sensitivity) is measured on the vertical axis while the model's false positive rate (lack of specificity) is on the horizontal axis. If the AUC for an ROC is 0.5, i.e. the ROC is coincident with a 45° line within the unit square, the model's predictive power is no better than a coin toss. Consequently, a model with an AUC of 1.0 (ROC corresponding to a single point in the upper-left corner) would be a 'perfect' predictor by the AUC measure.

Using a minimum threshold of 0.5 for classifying a case as 'recession,' ANN performs the worst (C-Statistic of 0.494) by the AUC. All the other methods have AUCs ranging from 0.616 (tree) to 0.778 (forest). The predictor derived from averaging the predicted probabilities of all six models is about as good as the Random Forest algorithm with a C-statistic of 0.777. Thus, tree-based models perform quite well out-of-sample. However, the AUC is not always a great measure of predictive quality.

Sensitivity and specificity matter in different ways for policy. A more specific model reduces the risk of misidentifying recessions, though it may miss some recessions. A more sensitive model reduces the risk of missing recessions though it may misidentify more periods of growth as recessions. The Random Forest performs quite well in terms of specificity (0.984) but badly in terms of sensitivity (0.101). The Bagging algorithm on the other hand does quite well in sensitivity (0.827) but poorly on specificity (0.039). Algorithms with high sensitivity are particularly useful if the goal is to identify high-risk countries for recession. By this measure, the Bagging model performs the best.

Mistakenly predicting recessions when there are none can be a costly policy response through say an increase in government spending. In that case, positive

---

[3]The story goes that the concept of ROC originated in WWII when the British began using radar to identify enemy aircraft. As radar sensitivity was increased, they were able to identify more enemy aircraft, but also more flocks of birds. Thus, increasing sensitivity increased the error rate, i.e. the identifications became less specific. The ROC curve takes this problem into account.

predictive value (PPV[4])—the probability that countries that are predicted to be in recessions in the next 5 years actually do experience recessions—may be a more relevant criterion for choosing a model. By this criterion, Random Forests perform the best (0.647), though the average is not too bad (0.611). Moreover, if policymakers are concerned about misidentifying countries with a higher recession risk then they could also look at negative predictive value. Negative predictive value (NPV[5]) is the probability that an algorithm predicts growth in countries that are truly growing. Here we see the single tree (0.808) and the Logit algorithm (0.804) perform the best, along with the average (0.802).

What can we conclude from all these results? Depending on the predictive criteria, the highest scoring models are Random Forest and Bagging models (except in the case of the negative predicted value). Overall then it appears that some form of bootstrapping and averaging of model parameters for a given method may improve predictive performance. Recent attempts at predicting recessions (Dopke et al. 2017; Berge 2015) confirm this intuition. However, our other distinction from these papers is quite critical. Based on our analysis of the comparison of different prediction algorithms along different predictive quality criteria suggests that prediction quality is sensitive to the *choice* of prediction quality criteria. We suggest that there is no silver bullet predictive algorithm. Therefore, policy makers should be very clear on the kind of prediction error they want to avoid before picking a prediction algorithm. This choice has implications.

Rose and Spiegel's (2011) review paper on the causes of the 2008–2009 economic crisis evaluates potential competing causes for that recession. They find no robust cause for that recession and suggest that the search for an early warning system for recessions is futile. Our algorithmic approach allows us to take a slightly more nuanced view on recessionary early warning systems. For example, say a policy maker is concerned about mistakenly implementing an expensive policy to preempt a recession. She would then rather miss a few recessions than implement an anti-recessionary policy that ex post turns out to be inappropriate. Such a policymaker would care more about specificity than sensitivity and could use the Random Forest technique that has the highest score on specificity (0.984) relative to all the other algorithms.[6] Our results would also let her know that she would be missing many recessions since the sensitivity score for the Random Forest is 0.1. However, the PPV of 0.647 (the highest score among the different algorithms)

---

[4]PPV = number of actual recessions/predicted number of recessions.

[5]NPV = number of actual growth events/predicted number of growth events.

[6]These kinds of choices have implications beyond economics. Recently doctors have suggested skipping some post-surgery preventive chemical/radiation treatment for breast cancer. In the past, then, doctors had been more concerned about sensitivity, missing post-surgery cancer flare ups. They weighed the health costs of treatment when there is likely to be no post-surgery (sensitivity errors) cancer against the health costs of missing some post-surgery cancers (specificity errors). They now seem to favor specificity over sensitivity in the detection of post-surgery cancers and have changed their recommendations accordingly (http://www.npr.org/sections/health-shots/2017/06/29/534882955/tumor-test-helps-identify-which-breast-cancers-dont-require-extra-treatment).

might help guide her in not giving a higher weight to sensitivity in her decision process. Of course, if she cared more about sensitivity she would choose the Bagging technique with a score of 0.827. Moreover, since ANNs perform uniformly poorly out-of-sample relative to the tree based models she would have access to a ranking of policy levers depending on her choice of predictive quality. We therefore turn to this ranking capability next. As we will show below in the next section, this process helps us add to the conversation on the nature of recessions as well as inform actionable policy decisions.

6.2   Variable Importance and Partial Dependence Plots: What Do We Learn?

Machine Learning algorithms rank predictive variables and develop partial dependence plots to provide insights into the sorts of variables that *may* cause recessions. We report our results on ranking in this section. We then view our results through two lenses. First, we suggest that our results can shed light on theoretical models of recessions. Second, we suggest that given a particular criterion for predictive quality such as specificity, ranking gives the policy maker a *validated* sense of which policy levers are likely to work the best.

6.2.1   The First Lens: Implications for Modeling Recessions Theoretically

Theoretical models of economic recessions roughly fall into two categories, real business cycle (RBC) and Keynesian approaches. In RBC approaches, typically the model always converges to a balanced growth (for example in the Ramsey model) unless there is some kind of a shock to the system from say some new technological invention that influences the production function (Kydland and Prescott 1982) or even increases in government spending (Baxter and King 1993) or the labor force (Romer 2001, pp. 173). Notice here the emphasis on shocks to the production function—this implies that recessions are proximately caused by sudden changes in capital, labor, or technology. The emphasis here therefore is on real supply side shocks, wages and prices adjust immediately to negate any demand side effects. The Keynesian approach on the other hand emphasizes that wage-price stickiness implies economic shocks cause changes to aggregate demand at any price level. This translates into a shock in the money market which then affects real output (Romer 2001, pp. 217). In short, unlike in RBC models, wage-price stickiness increases the salience of the monetary sector and variables like inflation, exchange rates, and interest rates in explaining recessions. In terms of predictors then, if RBC is a better theoretical model of recessions then real variables should be better predictors of recessions than monetary variables.

Our validated algorithms suggest that real variables may be better predictors of recessions than monetary variables. This is consistent with Berge's (2015) finding that real variables are better predictors of US economic downturns than monetary variables. Berge suggests that industrial production, monthly employment gains, initial jobless claims, and the change in the unemployment rate are better than purely monetary indicators like real money growth or the value of the trade weighted dollar at predicting recessions. Dopke et al. (2017) alternatively suggest that variables susceptible to monetary influences like nominal short-term interest rates and term spreads are the best predictors of recessions in Germany. Institutional variations between countries may be one possible source for these variations in findings.

The reader will recall that we do not have the exact same variables noted in the previous paragraph in our predictive models. However, we too have distinct categories of variables that capture real and monetary effects. In addition, we have institutional and socio-political variables that can serve as a proxy for the sort of institutional variations that may account for why monetary variables may predict recessions better in one country and real variables in another. For example, there may be institutional barriers to quick wage-price adjustments in Germany that are less salient in the US. In that case, though, when data is aggregated over time and countries, those institutional variables should have greater predictive salience than either the real or the monetary variables. Primed with these ideas, we turn to our results in Table 6.2.

Table 6.2 ranks variables that have the most important impact on the prediction for each of the tree models. We use the Gini impurity index to create this ranking.[7] We report the variable importance measures for each variable for the tree models, as well as the average importance for each variable's importance across all four Tree models. There is a wide variation in the predictive quality of the different tree methodologies. Therefore, we choose to aggregate the predictive information from the different methodologies and list the variables in order of average importance.[8]

We note first that with the exception of democracy the top ten predictors of recessions happen to be variables with a very "real" flavor. However, export price, which we argue is a monetary variable, is ranked as the 8th most important predictor. Inflation is the next monetary variable that enters into the predictive mix at rank 16. All the other monetary variables like interest rate spreads, money stock (LagMoneyGDP) and money growth rates (LagMoneyGrowth) appear even lower as predictive variables. Thus, while our results cannot be considered as definitive proof of the primacy of RBC over Keynesian approaches to understanding recessions it certainly seems that in terms of predictive out-of-sample fit (and subject to the variation in predictive quality discussed above) they seem to favor the RBC approach. This favor appears to be consistent across different countries since none of the variables that capture economic

---

[7]Gini impurity measures the probability that a randomly-selected element at any node, k, of a tree will be labeled incorrectly based on the decision rule at that node. Formally, $I_G(p_i) = \sum_{i=1}^{n_k} p_i(1 - p_i) = 1 - \sum_{i=1}^{n_k} p_i^2$.

[8]For example, lagged GDPpcGrowth reduces the Gini by 5.1% on average if it is removed from the predictive models $(3.5 + 4.05 + 7.4 + 5.5)/4 = 5.1$.

Table 6.2 Variable importance

Variable	Tree	Forest	Bagging	Boosting	Average
LagGDPpcGrowth	3.450	4.052	7.378	5.520	5.100
LagDemocracy	4.486	3.613	5.867	4.012	4.495
LagGDPpc	5.520	2.982	3.632	4.739	4.218
LagLifeExpectancy	4.667	3.213	3.662	4.869	4.103
LagPopulation	3.163	2.938	3.136	4.984	3.555
LagDependency	2.188	3.182	3.020	3.926	3.079
LagPopulationGrowth	1.938	3.039	2.922	4.162	3.016
LagExportPrices	1.640	2.793	3.055	4.435	2.981
LagInvestmentGDP	1.841	2.808	3.111	4.036	2.949
LagIndustryGDP	3.085	2.441	2.610	3.584	2.930
LagFDI_In_GDP	2.625	2.801	3.045	3.140	2.903
LagPhonesPC	3.691	3.015	2.454	2.406	2.891
LagRuralPopulationPCT	3.741	2.590	2.455	2.098	2.721
LagInflationCPI	2.839	2.309	2.340	3.169	2.664
LagExportsGDP	4.036	2.158	2.307	2.071	2.643
LagPrimCommodExports	1.807	2.685	2.849	3.072	2.603
LagImportPrices	1.052	2.484	2.377	4.077	2.497
LagTermsOfTrade	2.079	2.898	3.061	1.770	2.452
LagSecEnrollRatePCT	2.137	2.675	2.505	2.452	2.442
LagAidAssistGDP	2.574	2.432	2.656	1.624	2.322
LagConsumptionGDP	2.967	2.198	1.992	1.995	2.288
LagWithinInstab	1.219	2.594	2.848	2.247	2.227
LagMoneyGDP	2.091	2.534	2.115	2.148	2.222
LagTransparency	2.278	2.073	1.881	2.554	2.197
LagFemaleLaborForce	2.457	2.102	1.545	2.449	2.138
LagLaborForceParticipation	1.793	2.267	1.862	1.776	1.925
LagTradeGDP	3.602	1.830	1.271	0.887	1.897
LagMilitaryGDP	2.309	2.166	1.598	1.513	1.897
LagCredibility	1.832	2.191	1.923	1.302	1.812
LagGini	1.870	2.114	1.779	1.408	1.793
LagImportsGDP	2.961	1.786	1.331	0.846	1.731
LagInterestSpread	2.004	1.881	1.441	1.545	1.718
LagLendingInterestRate	2.430	1.838	1.442	0.866	1.644
LagFDI_Out_GDP	1.627	2.171	1.426	1.197	1.605
LagSchoolExpendGDP	1.665	1.671	1.464	1.398	1.549
LagNetGovernmentGDP	1.931	1.694	1.423	0.907	1.489
LagMoneyGrowth	1.546	1.789	1.745	0.505	1.396
LagTerror	1.560	1.575	1.257	1.181	1.393
LagRealInterestRate	1.079	1.691	1.574	0.777	1.280
LagProtest	0.772	1.575	1.449	1.192	1.247
LagRegimeInstab	0.754	1.383	1.436	0.800	1.094
LagExecutive	0.201	0.588	0.319	0.236	0.336

(continued)

Table 6.2 (continued)

Variable	Tree	Forest	Bagging	Boosting	Average
LagSystem	0.121	0.569	0.135	0.100	0.231
LagRegime	0.374	0.351	0.155	0.000	0.220
LagNonethnicConflict	0.000	0.119	0.073	0.023	0.054
LagEthnicConflict	0.000	0.138	0.070	0.000	0.052

institutional variations—e.g. credibility or transparency—appears to trump the variables that capture real economic activity. However, it is clear that monetary variables are worse at predicting recessions than real variables. The question however remains, what is a policy maker to do with all this?

6.2.2   The Second Lens: A Policy Maker and a Data Scientist Walk into a Bar

How could a policy maker use the capabilities of Machine Learning algorithms to smooth business cycles? We suggest a multi-step process. The first step must start with a sense of what kind of prediction errors a policy maker wants to avoid. For example, say a policy maker decides that she would rather miss a few recessions than implement an anti-recessionary policy that ex post turns out to be inappropriate. In other words, she would like to avoid specificity errors. She would note from Table 6.1 that the Random Forest algorithm with a score of 0.984 reduces specificity errors the most. That is, there is a 98.4% chance that the algorithm will predict an expansion when there is actually an expansion (specificity). She will know that there is 64.7% chance (PPV) that there is actually a recession given that the model predicts a recession. Nevertheless, she will also know that there is only a 10% chance of predicting recessions when there is actually a recession (sensitivity). She might also note that the Random Forests have higher scores across more predictive quality criteria than the other algorithms. She may then conclude that maybe the best she *can* do is avoid specificity errors with Random Forest algorithms. We are not saying that a policy maker should choose the Random Forest technique. Rather, we are outlining a how a policy may potentially use results like those shown in Table 6.1 to help inform a choice of predictive technique. Thus, we suggest that the looking at prediction quality over different criteria can sharpen the policy maker's thinking on what sorts of category identification mistakes they may want to avoid. This may reduce some of the reliance on "common sense" (it is unclear to us just how common "common" sense is!). Rodrik (2006) seems to want policy makers to fall back on in identifying expansionary constraints. Of course, the policy maker may just throw her hands up in the air and choose the average of predictions over the various approaches, given that the tree approaches are overall less prone to error.

Say the policy maker has decided on using the Random Forest algorithm. She can then use this algorithm to rank variables by contribution to out-of-sample fit, i.e. predictive salience. Table 6.3 does this for the top ten predictors.

Table 6.3 Top 10 policy levers chosen by the Random Forest algorithm

Variable	Forest
LagGDPpcGrowth	4.052
LagDemocracy	3.613
LagLifeExpectancy	3.213
LagDependency	3.182
LagPopulationGrowth	3.039
LagPhonesPC	3.015
LagGDPpc	2.982
LagPopulation	2.938
LagTermsOfTrade	2.898
LagInvestmentGDP	2.808

Variables that are more important are better candidates as policy levers because they have a higher effect on the likelihood of a recession.[9] How do they affect the likelihood of a recession? PDPs answer this question. The outer tick marks in the PDPs mark fixed increments of the value of each variable, while the inner tick marks indicate the cutoffs for each decile of the distribution of the variable. The vertical axis measures an odds ratio for the likelihood of a recession. These plots help visualize how a change in a covariate affects the likelihood of a recession in the learning sample.[10] We report the plots over three different algorithms, Bagging, Boosting, and Random Forests.

Countries with a high level of economic growth are less likely to have a recession (Fig. 6.2). The Random Forest algorithm chooses past GDP growth as the most important predictor of a recession. The policy maker therefore should prioritize policies that sustain growth to avoid recessions. Based on our analysis in the previous chapter this would mean policies that for example, encourage technology diffusion, increase both domestic capital investment and an inflow of foreign direct investment, and invest in healthcare. Notice this goes beyond saying that recessions should be fought with fiscal policy—we can pinpoint what *kinds* of fiscal policy are more likely to be effective. In fact, we notice that an improvement in life expectancy (Fig. 6.3)

---

[9]This table differs from the Table 6.2 because there the variables were ranked according to the average percentage reduction in Gini. Here we only use the Random Forest.

[10]Partial dependence plots display the marginal effect of variable x_k conditional on the observed values of all of the other variables, $(x_{1,-k}, x_{2,-k}, \ldots x_{n,-k})$ Specifically, it plots the graph of the function:

$$\hat{f}(x) = \frac{1}{n} \sum_{i=1}^{n} f(x_k, x_{i,-k}),$$

where the summand, $f(x_k, x_{i,-k})$, is defined by:

$$f(x) = \ln [p_1(x)] - \frac{1}{C} \sum_{s=1}^{C} \ln [p_s(x)].$$

Here, s indexes the classes of the target variable and C is the total number of classes. So, if an increase in a given factor, x_k, increases the probability of state failure ($s = 1$), then the value of the function plotted by the PDP also increases.

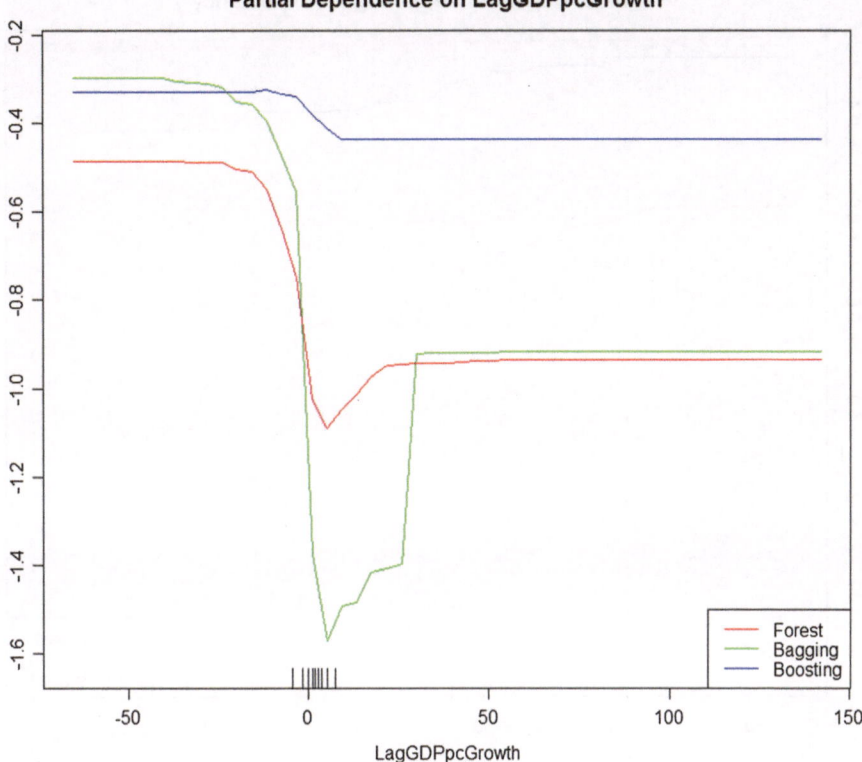

Fig. 6.2 Marginal effect of past growth on the likelihood of a recession

reduces the likelihood of a recession, doubly confirming that investment in health care builds immunity to recessions. Nevertheless, this is not a permanent vaccination. The highest levels of life expectancy increase the risk of a recession.

On the other hand, improvements in communication technology also have a peculiar effect on the likelihood of recessions (Fig. 6.4). As more phones are used per capita, the likelihood of recessions goes down. However, as phone penetration increases in the population the likelihood of recessions goes up! Odd as this may be, it highlights another feature of these PDPs that are not available with parametric point estimation techniques. Where a country is in the range of values for per capita phones determines whether a policy maker should increase investments in communication technology or not. In fact, the reader will recall how small increases in per capita phones shifted countries from low to high growth. Together with our finding here, the policy maker can suggest that the biggest bang for the buck for investing in communication technology for increasing growth and preventing recessions may come at relatively low levels of investment. Further, as dependency ratios rise (Fig. 6.5), initially, the likelihood of recessions goes down and then up. This tracks the effect of dependency ratios on economic growth suggesting that increases in dependency may initially increase growth while decreasing the risk of recessions. At this point, we do not know why this may be. Nevertheless, our findings are consistent. This curious finding given the predictive

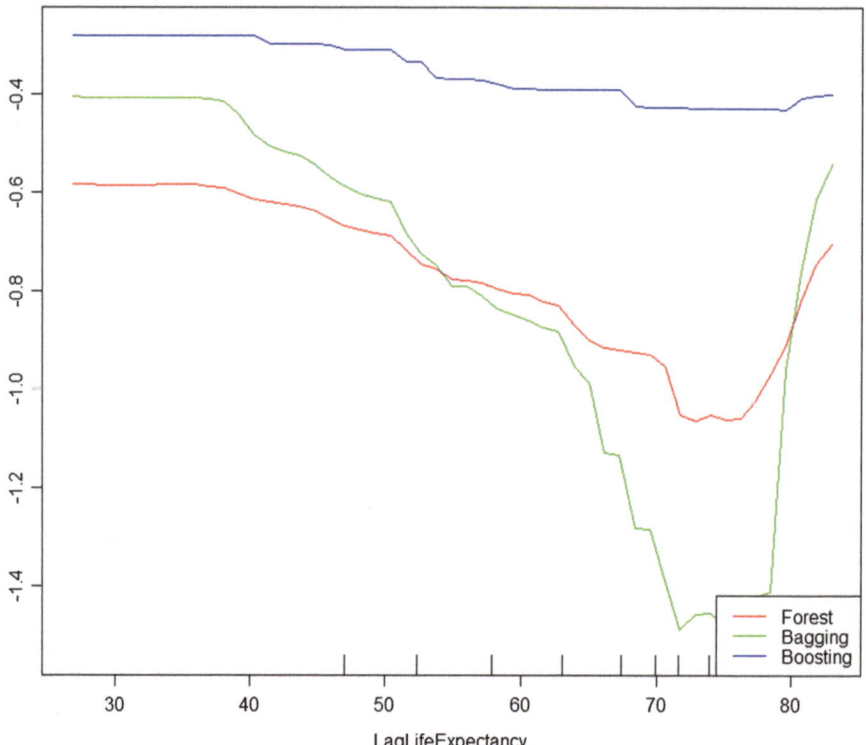

Fig. 6.3 Marginal effect of life expectancy on the likelihood of a recession

salience of dependency ratios for both growth and recessions suggests a closer look at the role of dependency in the growth literature. For our purposes here, this finding once again highlights the point that the policy maker needs to be concerned about where a country lies within the range of values for a policy lever (Figs. 6.4 and 6.5).

Figure 6.6 shows that initial increases in democracy decreases the likelihood of a recession. In the previous chapter, we showed that initial increases in democracy helps economic growth. Once again, it is heartening to find such consistent results. It is interesting to note that small steps toward political emancipation can have large effects on persistent economic growth. In fact this may help explain, for example, economic growth without huge recessions in much of the ASEAN region. Most of these countries achieved their biggest and most persistent (recession free) growth spurts without being paragons of Jeffersonian democracy (though there was some degree of political liberalization).

Population growth increases the risk of recessions (Fig. 6.7). This too tracks the effect of population growth rates on economic growth noted in the previous chapter. Higher population growth rates reduce economic growth and increase the risk of recession. This result highlights a certain Malthusianism. Nevertheless, we can

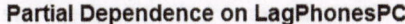

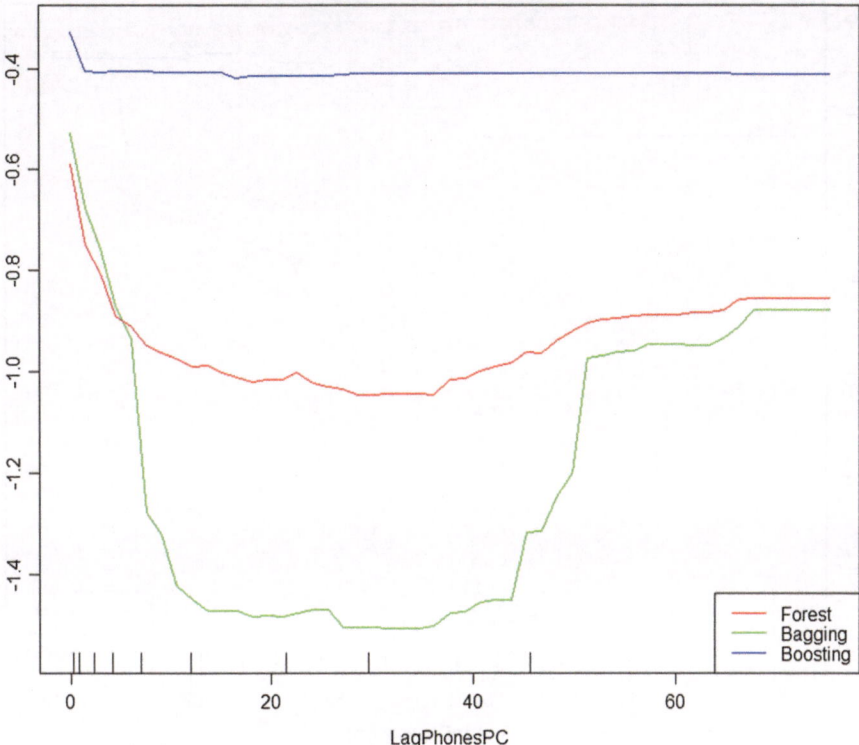

Fig. 6.4 Marginal effect of phones per capita on the likelihood of a recession

make the same argument we made before. We may not need to be quite as concerned about population growth because population growth is endogenous to economic growth. Thus, investments in healthcare that increase life expectancy increases economic growth and reduces the likelihood of recessions. This steady increase in economic well being may well increase the opportunity cost of having children and reduce population growth rates. Thus, health care investment may be a more salient policy lever than investing in, say, population curbs, like in China.

A policy maker should also be forewarned about increasing investment (physical capital) to reduce the risk of recession. While the effect of investment on economic growth is straightforward, increasing investment can have a non-linear effect on the likelihood of recessions. However, recall again that increasing investment in the lower range has the biggest effect on economic growth, shifting a country from a low to a high growth equilibrium. The result reported in Fig. 6.8 shows that initial increases in investment may decrease the risk of recession. Beyond a certain level of investment though the risk of recessions rise back up, though overall the trend is to have a slightly lower risk of recession as investment levels rise. This result may also point toward increased economic volatility with initial increases in investment.

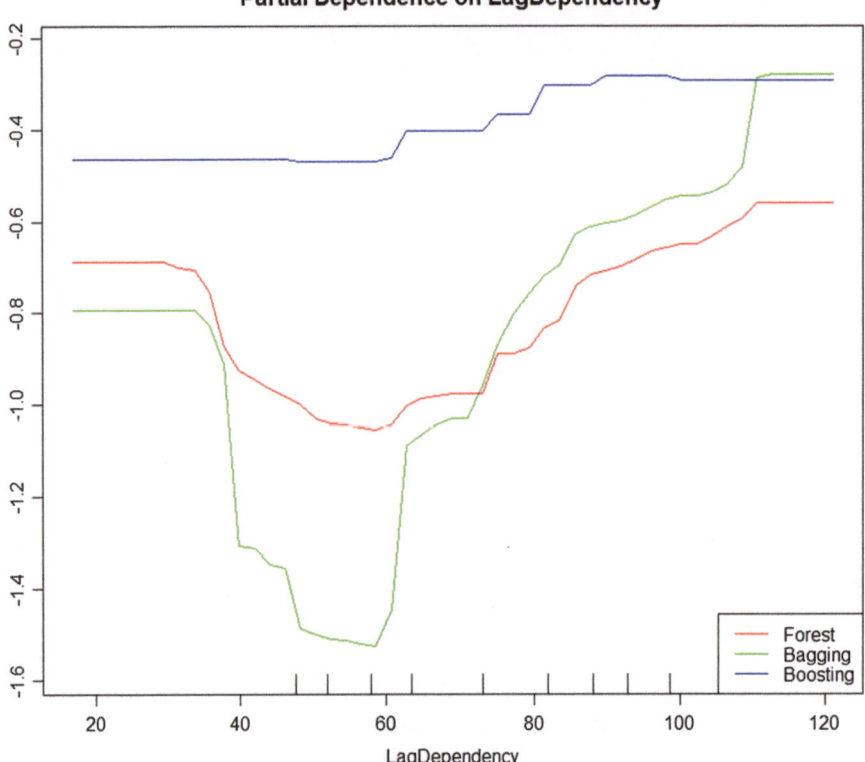

Fig. 6.5 Marginal effect of dependency on the likelihood of a recession

In the end, the policy maker will also note that real variables tend to be better predictors of the likelihood of recession for all the algorithms we have used. This suggests that fiscal policy is more effective than monetary policy in preventing recessions. Further, certain kinds of fiscal policy may be more salient for reducing the likelihood of recessions than others.

Coda

The discussion above suggests that Machine Learning algorithms can help evaluate the explanatory salience of different theoretical models of recession. Moreover, once a policy maker has decided what kind of errors to avoid, the same Machine Learning algorithms that informed the error avoidance decisions can also inform practical policy making. For example, lower ranked variables probably are less salient policy levers than higher ranked variables. Machine Learning algorithms thus provide guidance on what not to do! The policy maker would of course note that there is no policy silver bullet. In the case we discuss here, all the wonderful policy pre- scriptions are subject to the criticism that the focus on specificity implies that many recessions will be missed. Other error avoidance criteria may have other caveats. This

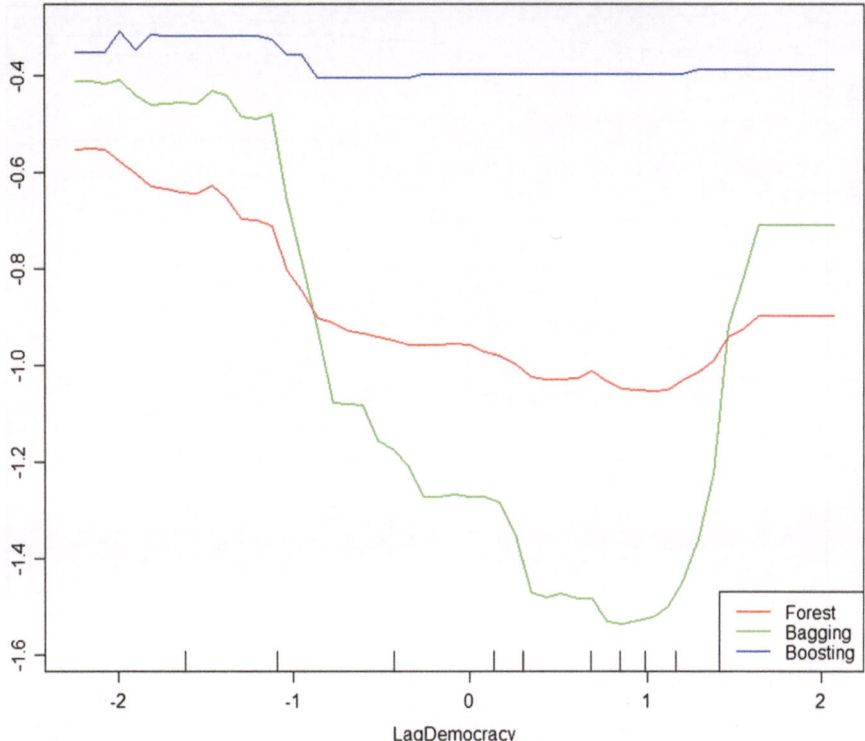

Fig. 6.6 Marginal effect of democracy on the likelihood of a recession

suggests policy makers should have a certain humility about their prescriptions—and a healthy dose of skepticism from the people affected by these policies. Nevertheless, there is some solace that the sorts of variables that appear to influence economic growth also seem to influence the likelihood of recessions (and irrespective of algorithm choice; notice the variables that are ranked by order of average importance in Table 6.1 relative to those ranked in Table 5.1). At one level, this suggests that economic growth probably ought not to be studied separately from GDP fluctuations. At another level, it tells the policy maker that the sorts of variables that influence economic growth may also prevent recessions. That consilience ought to count for something. Machine learning can therefore serve as a methodological framework for evaluating the relative salience of theory and policy through a predictive lens.

Takeaways

1. The prediction performance of algorithms differs widely depending on the type of prediction criteria. Policy makers and researchers need to be very sure of the kind of misclassification errors they want to avoid before using a criterion to pick the "best" algorithm.

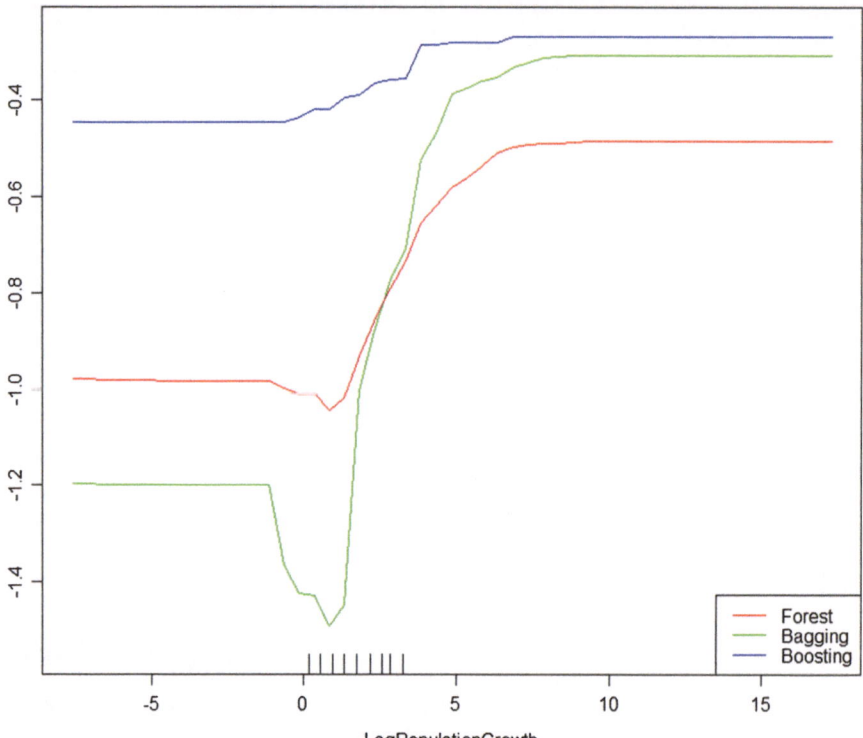

Fig. 6.7 Marginal effect of population growth rates on the likelihood of recessions

2. Researchers and policy makers should not depend on any one measure of predictive quality.
3. Policy levers may have a non-linear influence on the likelihood of recessions. Thus if a country falls within a certain range of a variable, increasing that variable may increase the risk of a recession, while in other ranges it may reduce the risk of a recession. In short, policies to temper GDP fluctuations may need to be very country specific.
4. Across time and space, real economic variables appear to be better predictors of recessions than either monetary or institutional variables. Thus, fiscal policy may be better at preventing recessions than monetary policy.
5. The sorts of variables that influence economic growth also influence the likelihood of a recession. This suggest that economic growth probably should not be studied separately from recessions.

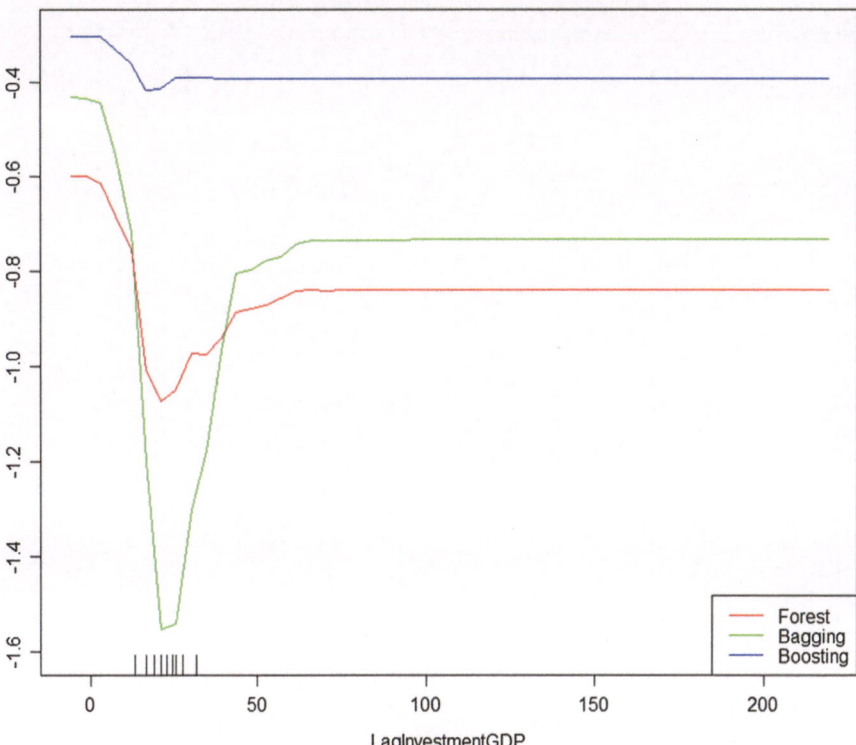

Fig. 6.8 Marginal effect of investment on the likelihood of a recession

References

Baxter, M., & King, R. G. (1993). Fiscal policy in general equilibrium. *American Economic Review, 83*, 315–334.

Berge, T. J. (2015). Predicting recessions with leading indicators: Model averaging and selection over the business cycle. *Journal of Forecasting, 34*, 455–471.

Dopke, J., Fritsche, U., & Pierdzioch, C. (2017, October and December). Predicting recessions with boosted regression trees. *International Journal of Forecasting, 33*(4), 745–759.

Kydland, F. E., & Prescott, E. C. (1982). Time to build and aggregate fluctuations. *Econometrica, 50*, 1345–1370.

Rodrik, D. (2006). Goodbye Washington Consensus, Hello Washington Confusion? A review of the world banks "Economic growth in the 1990's: Learning From a Decade of Reform". *Journal of Economic Literature, XLIV*, 973–987.

Romer, D. (2001). *Advanced macroeconomics*. New York: McGraw Hill Education.

Rose, A. K., & Spiegel, M. M. (2011). Cross country causes and consequences of the crisis: An update. *European Economic Review, 55*, 309–324.

Epilogue

We have highlighted relevant takeaways in many of the preceding chapter. In this epilogue, we organize these takeaways into three categories—methodological, contributions to theory, and contributions to policy analysis. We hope this organization may be useful for readers who may want to use this book for different purposes.

Methodological Takeaways

1. Some of the variables can be deconstructed to non-overlapping dimensions using EFA. This process also controls for unsystematic measurement errors.
2. Machine Learning fills in missing data with validated imputation techniques.
3. Machine Learning techniques help avoid researcher bias while giving the researcher a sense of the generalizability of a set of results.
4. The prediction performance of algorithms differs widely depending on the type of prediction criteria for classifying categorical variables. Policy makers and researchers need to be very sure of the kind of misclassification errors they want to avoid before using a criterion to pick the "best" algorithm.
5. Machine learning algorithms can be useful for *eliminating* policy levers that do not work.
6. Machine learning can serve as a methodological *framework* for evaluating the relative salience of theory and policy through a predictive lens.

Theoretical Takeaways

1. We include data that mirrors Sala-i-Martin's (1997) list of robust covariates from EBA analysis. These variables originate from the major strands of theories of growth. Therefore, they represent major growth theories quite comprehensively.
2. The robust correlates of economic growth suggested by EBA (Sala-i-Martin 1997) are not very practical predictors of a country's economic growth. Nevertheless, we learn something about economic growth by eliminating competing theoretical models of growth.

© The Author(s) 2017

A. Basuchoudhary et al., *Machine-learning Techniques in Economics*,
SpringerBriefs in Economics, https://doi.org/10.1007/978-3-319-69014-8

3. The magnitudes of the prediction errors do not seem to be influenced by some of the usual problems (Temple 1999) in the empirical growth literature.
4. A good theoretical model should also be predictive. To the extent that a model based on current growth theory has prediction problems suggests an emphasis on understanding economic growth beyond the usual suspects.
5. Multiplicity of growth theories is less of a problem than lack of completeness in current growth theory.
6. It is important for future research to focus on cultural and other intergenerational information transfer mechanisms to get a more complete understanding of growth since geography, history, and institutions (other than democracy) appear to have little predictive value in terms of out-of-sample fit.
7. Many of the variables that EBA suggests as robust correlates of economic growth are not good predictors. This suggests that EBA is not parsimonious enough in helping focus on potential causal factors of economic growth.
8. The variables that do predict growth well suggest that the process of growth is not a smooth secular increase—rather each country faces an equilibrium shift from low to high growth. Thus, poverty traps are a very real possibility.
9. The sorts of variables that influence economic growth also influence the likelihood of a recession. This suggest that economic growth probably should not be studied separately from recessions.

Policy Takeaways

1. Trade deficits financed by FDI increases growth as do investments in health care and communication. Moreover, policy makers should be cautious about potential terms of trade shocks. Particularly for developing countries, these results suggest capital account liberalization with restraints in the current account.
2. Democracy is the only predictive institutional variable.
3. In the absence of variables that capture intergenerational cultural transmission mechanisms, endogenous growth models appear to explain the binding constraints on growth common to all countries. This suggests that policy should focus on easing these constraints rather than relying on a policy maker's gut instinct.
4. Policy levers may influence the likelihood of recessions non-linearly. Thus, if a country falls within a certain range of a variable, increasing that variable may increase the risk of a recession, while in other ranges it may reduce the risk of a recession. In short, policies to temper GDP fluctuations may need to be very country specific.
5. Across time and space real economic variables appear to be better predictors of recessions than either monetary or institutional variables. Thus, fiscal policy may be better at preventing recessions than monetary policy.

Appendix: R Codes and Notes

In this Appendix, we provide some annotated code so that researchers seeking to replicate our project (or apply the methods in this paper to their own ideas) will have a roadmap. Note that we have trained all of our models in R, but we have used other programs for some of the preprocessing and structuring of the data. For the purpose of brevity, we focus on imputing the data using Random Forests; training the models; calculating the various accuracy measures; and drawing the PDPs. We assume an elementary proficiency with loading data and packages in R throughout. The codes are italicized to separate them from the notes. Periods at the end of each code block are inserted for grammar and are not part of the code.

Imputing and Processing the Data

First, we load the data from the file we have used to merge the data:

library(foreign);
GrowthData<-read.dta("GrowthData.dta").

We will also need to make sure that *R* reads all of the factor variables as factors instead of numerical/ordinal variables:

GrowthData$PartyExclusion <-as.factor(GrowthData$PartyExclusion)
GrowthData$Executive<-as.factor(GrowthData$Executive)
GrowthData$Regime<-as.factor(GrowthData$Regime)
GrowthData$PartyCoalitions<-as.factor(GrowthData$PartyCoalitions)
GrowthData$ParliamentRespons<-as.factor(GrowthData$ParliamentRespons)
GrowthData$System<-as.factor(GrowthData$System).

Now, we begin with the imputation using the randomForest package. We set the random seed to "8976" in order to ensure that the dataset will be replicable.

© The Author(s) 2017
A. Basuchoudhary et al., *Machine-learning Techniques in Economics*,
SpringerBriefs in Economics, https://doi.org/10.1007/978-3-319-69014-8

```
library(randomForest)
set.seed(8976)
GrowthData.rfImpute<-rfImpute(Conflict                                    ~
    GDPpc+ConsumptionGDP+InvestmentGDP+NetGovernmentGDP+Military-
    GDP+ImportsGDP+IndustryGDP+AidAssistGDP+TradeGDP+ExportsGDP
    + PrimCommodExports + TermsOfTrade + ExportPrices + ImportPrices +
    FDI_In_GDP + FDI_Out_GDP + Population + PopulationGrowth +
    LifeExpectancy    +    Dependency    +    LaborForceParticipation    +
    FemaleLaborForce    +    LifeExpectancy    +    SecEnrollRatePCT    +
    SchoolExpendGDP + RuralPopulationPCT + MoneyGDP + MoneyGrowth +
    InflationCPI + RealInterestRate + LendingInterestRate + InterestSpread +
    PhonesPC + Gini + GovStability + InvestmentProf + InternalConflict +
    ExternalConflict   +   Corruption   +   EthnicTensions   +   DemAcct   +
    BureaucraticQual   +   EthnicPolariz   +   ReligiousPolariz   +   LegFrac   +
    GovPolarization   +   ExecYrsOffice   +   ChangeInVetoes   +   GovHerfindahl   +
    Checks + LIEC + EIEC + Fraud + Polity2 + RegDurability + Assassinations
    + Strikes + GuerrillaWar + GovCrises + Purges + Riots + Revolutions +
    Demonstrations + Coups + CabinetChgs + ExecutiveChgs + LegElections +
    TerrorAttacks + Fatalities + Injuries + PropertyDamage + System + Executive
    + Regime + PartyExclusion + PartyCoalitions + ParliamentRespons + Year,
        data = GrowthData, ntree = 500, iter = 20, maxnodes = 128, nodesize =
    10, localImp = TRUE).
```

Now that we have our imputed data, we will expand the factor variables into sets of mutually-exclusive dummy variables that we can use in the parametric models like Logit and neural networks. This step uses the "class.ind" function from the "nnet" package.

```
library(nnet)
i.System<-class.ind(GrowthData.rfImpute$System)
i.PartyExcl <-class.ind(GrowthData.rfImpute$PartyExclusion)
i.Executive <-class.ind(GrowthData.rfImpute$Executive)
i.Regime <-class.ind(GrowthData.rfImpute$Regime)
i.PartyCoal <-class.ind(GrowthData.rfImpute$PartyCoalitions)
i.ParlResp <-class.ind(GrowthData.rfImpute$ParliamentRespons)
```

We will also give these dummy variables names based on the variable name and factor labels.

```
colnames(i.System)<-paste("Sys", colnames(i.System), sep = "_")
colnames(i.PartyExcl)<-paste("Excl",colnames(i.PartyExcl),sep = "_")
colnames(i.Executive)<-paste("Exec",colnames(i.Executive),sep = "_")
colnames(i.Regime)<-paste("Reg", colnames(i.Regime), sep = "_")
colnames(i.PartyCoal)<-paste("Coal",colnames(i.PartyCoal),sep = "_")
colnames(i.ParlResp)<-paste("Resp", colnames(i.ParlResp),sep = "_").
```

We will combine the new dummy variable with the imputed dataframe:

GrowthData.rfImpute<-cbind(GrowthData.rfImpute, i.System, i.PartyExcl, i.Executive, i.Regime, i.PartyCoal, i.ParlResp)

and also combine the country codes, target variable (which we did *not* impute), and ethnic/nonethnic conflict disaggregation of the conflict variable used for the imputation.

GrowthData.rfImpute<-cbind(GrowthData$wbcode, GrowthData$GDPpcGrowth, GrowthData$EthnicConflict, GrowthData$NonethnicConflict).

At this point we generated five-year moving averages and the five-year lags of these moving averages using standard data management functions. We have skipped these steps in this description and encourage researchers to use the most familiar tools they have to complete this structuring of the variables. Next, we will show the code we have used to divide our sample.

First, we will subset our data to include observations beginning in 1970 for which we have nonmissing values for our target variable and its five-year lag.

library(gdata)
GrowthData.Full <- subset(GrowthData.rfImpute, Year > 1970)
GrowthData.Full <- subset(GrowthData.Full, !is.na(GDPpcGrowthMA))
GrowthData.Full <- subset(GrowthData.Full, !is.na(LagGDPpcGrowth)).

The final step before we begin describing the code for training the models is to divide our sample into learning and test randomly by country.

set.seed(8976)
GrowthData.Full$rand <- runif(nrow(GrowthData.Full))
GrowthData.Full$randbar <- with(GrowthData.Full, ave(rand, wbcode, FUN = mean))
GrowthData.Full$randbarp <- rank(GrowthData.Full$randbar)/ length(GrowthData.Full$randbar)
GrowthData.LS <- subset(GrowthData.Full, randbarp <= 0.70)
GrowthData.TS <- subset(GrowthData.Full, randbarp > 0.70).

Now, the data are ready to train the models.

Training the Models

In this section, we provide annotated code for training our learning models. In it, we will use several packages: rpart, randomForest, nnet, gbm, and adabag will build our models; rpart.plot may also be useful for users who want to generate more visually pleasing diagrams of the CART model. We load the required R libraries:

library(rpart); library(rpart.plot); library(randomForest); library(nnet); library (pROC); library(gbm); library(adabag).

For simplicity, we will attach our learning sample to R's search path, and define our growth model formula.

attach(GrowthData.LS)
GrowthModelFormula <- formula(GDPpcGrowthMA ~ LagGDPpcGrowth + LagGDPpc + LagConsumptionGDP + LagInvestmentGDP + LagNetGovernmentGDP + LagMilitaryGDP + LagImportsGDP + LagIndustryGDP + LagAidAssistGDP + LagTradeGDP + LagExportsGDP + LagPrimCommodExports + LagTermsOfTrade + LagExportPrices + LagImportPrices + LagFDI_In_GDP + LagFDI_Out_GDP + LagPopulation + LagPopulationGrowth + LagLifeExpectancy + LagDependency + LagLaborForceParticipation + LagFemaleLaborForce + LagLifeExpectancy + LagSecEnrollRatePCT + LagSchoolExpendGDP + LagRuralPopulationPCT + LagMoneyGDP + LagMoneyGrowth + LagInflationCPI + LagRealInterestRate + LagLendingInterestRate + LagInterestSpread + LagPhonesPC + LagGini + LagEthnicConflict + LagNonethnicConflict + LagDemocracy + LagTransparency + LagCredibility + LagWithinInstab + LagRegimeInstab + LagTerror + LagProtest + LagSys_Parliamentary + LagSys_AssemblyElectedPresident + LagSys_Presidential + LagExec_Military + LagExec_Monarch + LagExec_Other + LagExec_Premier + LagExec_Premier + LagExec_President + LagReg_Civilian + LagReg_Military + LagReg_MilitaryCivilian + LagReg_Other)
RecessionModelFormula <- formula(Recession5 ~ LagGDPpcGrowth + LagGDPpc + LagConsumptionGDP + LagInvestmentGDP + LagNetGovernmentGDP + LagMilitaryGDP + LagImportsGDP + LagIndustryGDP + LagAidAssistGDP + LagTradeGDP + LagExportsGDP + LagPrimCommodExports + LagTermsOfTrade + LagExportPrices + LagImportPrices + LagFDI_In_GDP + LagFDI_Out_GDP + LagPopulation + LagPopulationGrowth + LagLifeExpectancy + LagDependency + LagLaborForceParticipation + LagFemaleLaborForce + LagLifeExpectancy + LagSecEnrollRatePCT + LagSchoolExpendGDP + LagRuralPopulationPCT + LagMoneyGDP + LagMoneyGrowth + LagInflationCPI + LagRealInterestRate + LagLendingInterestRate + LagInterestSpread + LagPhonesPC + LagGini + LagEthnicConflict + LagNonethnicConflict + LagDemocracy + LagTransparency + LagCredibility + LagWithinInstab + LagRegimeInstab + LagTerror + LagProtest + LagSys_Parliamentary + LagSys_AssemblyElectedPresident + LagSys_Presidential + LagExec_Military + LagExec_Monarch + LagExec_Other + LagExec_Premier + LagExec_Premier + LagExec_President + LagReg_Civilian + LagReg_Military + LagReg_MilitaryCivilian + LagReg_Other.)

This will save us the trouble of specifying the data frame in each model.

We start by training the parametric models, starting with the simple linear regression for growth.

Growth.lm <- lm(GrowthModelFormula)

The corresponding (generalized) linear model for the binomial recession variable is the Logit model, which we implement using *glm*.

Recession.lm <- glm(RecessionModelFormula, family = "binomial")

We follow this with the neural net model using 20 hidden nodes (size = 20), 1200 maximum number of weights. For the growth model, we use linear output units (instead of logistic units, which make more sense for classification than regression).

set.seed(8976)
Growth.nnet <- nnet(GrowthModelFormula, size = 20, linout = T, MaxNWts = 1200)
Recession.nnet <- nnet(RecessionModelFormula, size = 20, MaxNWts = 1200).

Now, we will train our nonparametric tree models (single tree, Bagging, Boosting, and Random Forest). For the tree model, we set the controls so that the complexity parameter (minimum improvement for a new split) equals 0.001, and the minimum number of observations for a new split equals 10 [control = rpart. control(cp = 0.001, minsplit = 10)].

Growth.tree <- rpart(GrowthModelFormula, control = rpart.control(cp = 0.001, minsplit = 10))
Recession.tree <- rpart(RecessionModelFormula, control = rpart.control(cp = 0.001, minsplit = 10)).

We use the gbm function to create the Boosting models. The options we specify include the distribution (distribution = "gaussian" instead of "bernoulli" to match the continuous target variable); the number of trees (n.trees = 500); and the depth of interactions (interaction.depth = 3 to allow 3-way interactions, in contrast to a depth of 1 for a linear model).

set.seed(8976)
Growth.boost <- gbm(GrowthModelFormula, distribution = "gaussian", n.trees = 500, interaction.depth = 3).

For classifying recessions, we use the simpler "boosting" command from the *adabag* package.

Recession.boost <- boosting(RecessionModelFormula, control = rpart.control (cp = 0.001, maxdepth = 5, minsplit = 10), data = GrowthData.LS, mfinal = 25)

To build the bootstrap aggregating (Bagging) model, we actually use the *randomForest* command. The main difference between Bagging is that the Bagging model only randomizes (bootstraps) the sample; the Random Forest randomizes the variables under consideration at each split. Hence, the two are equivalent when the number of variables the model randomly selects to choose from for the split each node equals the total number varaibles in the model (mtry = 39). We also set the minimum number of observations needed to consider a split to 10 (nodesize = 10); and the sample size for each bootstrap draw to be 3325 (about 73% of our learning sample, sampsize = 3325).

```
set.seed(8976)
Growth.bag <- randomForest(GrowthModelFormula, mtry = 39, nodesize =
    10, sampsize = 3325)
Recession.bag <- randomForest(RecessionModelFormula, mtry = 39, nodesize =
    10, sampsize = 3325).
```

The Random Forest model uses nearly the same arguments as the Bagging model, except that set the number of variables to select at each node to three (mtry = 3), and we do not specify a user-defined sample size for each tree.

```
set.seed(8976)
Growth.rf <- randomForest(GrowthModelFormula, mtry = 3, nodesize = 10)
Recession.rf <- randomForest(RecessionModelFormula, mtry = 3, nodesize = 10).
```

Evaluating Predictive Quality

In this section, we describe the code for assessing the accuracy of the models. To conserve space, we have omitted some of the code that combines and organizes these predictive measures into publishable tables. We begin by extracting the predicted values for growth. For each model, we create a predicted value. We also calculate the average of the predictions of the various models. For comparison purposes our chapter has presented prediction accuracy for both the learning and test samples. Here we focus on the test samples since these better measure true *predictive* power, whereas the corresponding measures for the learning sample better measure the *descriptive* power.

```
GrowthData.TS$GDPpcGrowthMA.lm   <-   predict(Growth.lm,   newdata   =
    GrowthData.TS)
GrowthData.TS$GDPpcGrowthMA.nnet  <-  predict(Growth.nnet,  newdata  =
    GrowthData.TS)
GrowthData.TS$GDPpcGrowthMA.tree  <-  predict(Growth.tree,  newdata  =
    GrowthData.TS)
GrowthData.TS$GDPpcGrowthMA.bag   <-   predict(Growth.bag,   newdata   =
    GrowthData.TS)
GrowthData.TS$GDPpcGrowthMA.boost <- predict(Growth.boost, newdata =
    GrowthData.TS, n.trees = 500)
GrowthData.TS$GDPpcGrowthMA.rf    <-    predict(Growth.rf,   newdata   =
    GrowthData.TS)
```

GrowthData.TS$GDPpcGrowthMA.ave <- (GrowthData.TS$GDPpcGrowthMA.lm + GrowthData.TS$GDPpcGrowthMA.bag + GrowthData.TS$GDPpcGrowthMA. boost + GrowthData.TS$GDPpcGrowthMA.rf)/4.

Our main measure of predictive quality for the growth models is the mean squared error. We calculate the overall MSE for the null model and also for each of the predictions, including the average of all of the predictors.

MSE.TS <- mean((GrowthData.TS$GDPpcGrowthMA - mean(GrowthData.TS $GDPpcGrowthMA))^2)

MSE.lm.TS <- mean((GrowthData.TS$GDPpcGrowthMA -GrowthData.TS $GDPpcGrowthMA.lm)^2)

MSE.nnet.TS <- mean((GrowthData.TS$GDPpcGrowthMA -GrowthData.TS $GDPpcGrowthMA.nnet)^2)

MSE.tree.TS <- mean((GrowthData.TS$GDPpcGrowthMA -GrowthData.TS $GDPpcGrowthMA.tree)^2)

MSE.bag.TS <- mean((GrowthData.TS$GDPpcGrowthMA -GrowthData.TS $GDPpcGrowthMA.bag)^2)

MSE.boost.TS <- mean((GrowthData.TS$GDPpcGrowthMA -GrowthData.TS $GDPpcGrowthMA.boost)^2)

MSE.rf.TS <- mean((GrowthData.TS$GDPpcGrowthMA-GrowthData.TS $GDPpcGrowthMA.rf)^2)

MSE.ave.TS <- mean((GrowthData.TS$GDPpcGrowthMA -GrowthData.TS $GDPpcGrowthMA.ave)^2).

We use several measures of predictive quality for the recession models. These include the sensitivity, specificity, positive predictive value, negative predictive value, area under the (receiver-operating characteristic) curve, and the overall error rate. Once again, we begin by calculating the predicted values of each model as well as an average of all of the models – for both the probability of recession and the predicted outcome (recession or not).

GrowthData.TS$treeProb<-predict(Recession.tree, type = "prob", newdata = GrowthData.TS)[,"Recession"]

GrowthData.TS$treeClass<-predict(Recession.tree, type = "class", newdata = GrowthData.TS)

GrowthData.TS$rfProb<-predict(Recession.rf, type = "prob", newdata = GrowthData.TS)[,"Recession"]

GrowthData.TS$rfClass<-predict(Recession.rf, type = "class", newdata = GrowthData.TS)

GrowthData.TS$bagProb<-predict(Recession.bag, type = "prob", newdata = GrowthData.TS)[,"Recession"]

GrowthData.TS$bagClass<-predict(Recession.bag, type = "class", newdata = GrowthData.TS)

GrowthData.TS$boostProb<-predict(Recession.boost, newdata = GrowthData. TS)$prob[,2]

*GrowthData.TS$boostClass<-as.factor(predict(Recession.boost,   newdata   =
GrowthData.TS)$class)*
*GrowthData.TS$lmProb<-predict(Recession.lm, newdata = GrowthData.TS, type
= "response")*
*GrowthData.TS$lmClass<-as.factor(ifelse(GrowthData.TS$lmProb   <   0.5,
"NoRecession", "Recession"))*
*GrowthData.TS$nnetProb<-as.vector(predict(Recession.nnet,   newdata   =
GrowthData.TS, type = "raw"))*
*GrowthData.TS$nnetClass<-as.factor(predict(Recession.nnet,   newdata   =
GrowthData.TS, type = "class"))*
*GrowthData.TS$averageProb<-(GrowthData.TS$treeProb   +   GrowthData.TS
$rfProb   +   GrowthData.TS$bagProb   +   GrowthData.TS$boostProb   +
GrowthData.TS$lmProb)/5*
*GrowthData.TS$averageClass<-ifelse(GrowthData.TS$averageProb   <   0.5,
"NoRecession", "Recession").*

Using the predicted classes, we build the confusion matrix for each model (for
both the test and learning samples).

Confusion.tree.TS<-with(GrowthData.TS, table(Recession5, treeClass))
Confusion.rf.TS<-with(GrowthData.TS, table(Recession5, rfClass))
Confusion.bag.TS<-with(GrowthData.TS, table(Recession5, bagClass))
Confusion.bag.TS <- cbind(Confusion.bag.TS[,2], Confusion.bag.TS[,1])
Confusion.boost.TS<-with(GrowthData.TS, table(Recession5, boostClass))
Confusion.boost.TS <- cbind(Confusion.boost.TS[,2], Confusion.boost.TS[,1])
Confusion.lm.TS<-with(GrowthData.TS, table(Recession5, lmClass))
Confusion.nnet.TS<-with(GrowthData.TS, table(Recession5, nnetClass))
Confusion.ave.TS<-with(GrowthData.TS, table(Recession5, averageClass)).

We calculate the sensitivity as the number of correctly-predicted recessions as a
proportion of the actual recessions.

*Sensitivity.tree.TS<-Confusion.tree.TS[2,2]/(Confusion.tree.TS[2,2] + Confusion.
tree.TS[2,1])*
*Sensitivity.rf.TS<-Confusion.rf.TS[2,2]/(Confusion.rf.TS[2,2] + Confusion.rf.TS
[2,1])*
*Sensitivity.bag.TS<-Confusion.bag.TS[2,2]/(Confusion.bag.TS[2,2] + Confusion.
bag.TS[2,1])*
*Sensitivity.boost.TS<-Confusion.boost.TS[2,2]/(Confusion.boost.TS[2,2] + Confusion.
boost.TS[2,1])*
*Sensitivity.lm.TS<-Confusion.lm.TS[2,2]/(Confusion.lm.TS[2,2] + Confusion.lm.
TS[2,1])*
*Sensitivity.nnet.TS<-Confusion.nnet.TS[2,2]/(Confusion.nnet.TS[2,2] + Confusion.
nnet.TS[2,1])*
*Sensitivity.ave.TS<-Confusion.ave.TS[2,2]/(Confusion.ave.TS[2,2] + Confusion.
ave.TS[2,1]).*

The specificity is the number of correctly-predicted *non*-recessions as a proportion of the number of actual non-recessions,

Specificity.tree.TS<-Confusion.tree.TS[1,1]/(Confusion.tree.TS[1,1] + Confusion. tree.TS[1,2])

Specificity.rf.TS<-Confusion.rf.TS[1,1]/(Confusion.rf.TS[1,1] + Confusion.rf.TS [1,2])

Specificity.bag.TS<-Confusion.bag.TS[1,1]/(Confusion.bag.TS[1,1] + Confusion. bag.TS[1,2])

Specificity.boost.TS<-Confusion.boost.TS[1,1]/(Confusion.boost.TS[1,1] + Confu-sion.boost.TS[1,2])

Specificity.lm.TS<-Confusion.lm.TS[1,1]/(Confusion.lm.TS[1,1] + Confusion.lm. TS[1,2])

Specificity.nnet.TS<-Confusion.nnet.TS[1,1]/(Confusion.nnet.TS[1,1] + Confusion. nnet.TS[1,2])

Specificity.ave.TS<-Confusion.ave.TS[1,1]/(Confusion.ave.TS[1,1] + Confusion. ave.TS[1,2]).

The positive predictive value (PPV) is the number of true positives as a proportion of the number of predicted positives.

PPV.tree.TS<-Confusion.tree.TS[2,2]/(Confusion.tree.TS[2,2] + Confusion.tree. TS[1,2])

PPV.rf.TS<-Confusion.rf.TS[2,2]/(Confusion.rf.TS[2,2] + Confusion.rf.TS[1,2])

PPV.bag.TS<-Confusion.bag.TS[2,2]/(Confusion.bag.TS[2,2] + Confusion.bag. TS[1,2])

PPV.boost.TS<-Confusion.boost.TS[2,2]/(Confusion.boost.TS[2,2] + Confusion. boost.TS[1,2])

PPV.lm.TS<-Confusion.lm.TS[2,2]/(Confusion.lm.TS[2,2] + Confusion.lm.TS [1,2])

PPV.nnet.TS<-Confusion.nnet.TS[2,2]/(Confusion.nnet.TS[2,2] + Confusion.nnet. TS[1,2])

PPV.ave.TS<-Confusion.ave.TS[2,2]/(Confusion.ave.TS[2,2] + Confusion.ave.TS [1,2]).

The negative predictive value is the number of true negatives as a proportion of the number of predicted negatives.

NPV.tree.TS<-Confusion.tree.TS[1,1]/(Confusion.tree.TS[1,1] + Confusion.tree. TS[2,1])

NPV.rf.TS<-Confusion.rf.TS[1,1]/(Confusion.rf.TS[1,1] + Confusion.rf.TS[2,1])

NPV.bag.TS<-Confusion.bag.TS[1,1]/(Confusion.bag.TS[1,1] + Confusion.bag. TS[2,1])

NPV.boost.TS<-Confusion.boost.TS[1,1]/(Confusion.boost.TS[1,1] + Confusion. boost.TS[2,1])

NPV.lm.TS<-Confusion.lm.TS[1,1]/(Confusion.lm.TS[1,1] + Confusion.lm.TS [2,1])

NPV.nnet.TS<-Confusion.nnet.TS[1,1]/(Confusion.nnet.TS[1,1] + Confusion.nnet.
TS[2,1])
NPV.ave.TS<-Confusion.ave.TS[1,1]/(Confusion.ave.TS[1,1] + Confusion.ave.TS
[2,1]).

The overall error rate (or redistribution rate) is the number of incorrect predictions as a proportion of the total sample.

Error.tree.TS<-(Confusion.tree.TS[1,2]   +   Confusion.tree.TS[2,1])/   nrow
(GrowthData.TS)
Error.rf.TS<-(Confusion.rf.TS[1,2] + Confusion.rf.TS[2,1])/ nrow(GrowthData.
TS)
Error.bag.TS<-(Confusion.bag.TS[1,2]   +   Confusion.bag.TS[2,1])/   nrow
(GrowthData.TS)
Error.boost.TS<-(Confusion.boost.TS[1,2]   +   Confusion.boost.TS[2,1])/   nrow
(GrowthData.TS)
Error.lm.TS<-(Confusion.lm.TS[1,2] + Confusion.lm.TS[2,1])/ nrow(GrowthData.
TS)
Error.nnet.TS<-(Confusion.nnet.TS[1,2]   +   Confusion.nnet.TS[2,1])/   nrow
(GrowthData.TS)
Error.ave.TS<-(Confusion.ave.TS[1,2]   +   Confusion.ave.TS[2,1])/   nrow
(GrowthData.TS).

We plot the receiver-operating characteristic (ROC) curves using the roc function in the pROC library.

library(pROC)
roc.tree.TS<-roc(GrowthData.TS$Recession5, GrowthData.TS$treeProb, smooth
= F, auc = T, plot = T, add = F, col = "brown", lwd = 1)
roc.rf.TS<-roc(GrowthData.TS$Recession5, GrowthData.TS$rfProb, smooth = F,
auc = T, plot = T, add = T, col = "red", lwd = 1)
roc.bag.TS<-roc(GrowthData.TS$Recession5, GrowthData.TS$bagProb, smooth
= F, auc = T, plot = T, add = T, col = "green", lwd = 1)
roc.boost.TS<-roc(GrowthData.TS$Recession5,   GrowthData.TS$boostProb,
smooth = F, auc = T, plot = T, add = T, col = "blue", lwd = 1)
roc.lm.TS<-roc(GrowthData.TS$Recession5, GrowthData.TS$lmProb, smooth
= F, auc = T, plot = T, add = T, col = "gold", lwd = 1)
roc.nnet.TS<-roc(GrowthData.TS$Recession5, GrowthData.TS$nnetProb, smooth
= F, auc = T, plot = T, add = T, col = "gray", lwd = 1)
roc.ave.TS<-roc(GrowthData.TS$Recession5,   GrowthData.TS$averageProb,
smooth = F, auc = T, plot = T, add = T, col = "black", lwd = 1).

The ROC objects store the area under the curve (AUC) as *$auc* part of the construction of the curve.

Variable Importance and Partial Dependence

Each package stores the variable importance slightly differently. In the *rpart* package it is stored as *$variable.importance*; in the *randomForest* package (which includes the Bagging model) it is *$importance[,column]*, where column equals 1 for regression (to get the % change in MSE) and 4 for classification (to get the mean decrease in the gini dispersion); the *gbm* package stores it as column 2 of its corresponding summary object; and the *adabag* package (which we use for the Boosting classification model) simply stores it as *$importance*).

For each importance measure, we calculate the importance in relative terms as a percentage of the total improvement for that model. We then take the simple average relative importance of the four tree models. The code for the growth regression tree models is:

Importance.tree<-100*as.data.frame(Growth.tree$variable.importance)/ sum (Growth.tree$variable.importance)

Importance.rf<-100*as.data.frame(Growth.rf$importance[,1])/ sum(Growth.rf $importance[,1])

Importance.bag<-as.data.frame(Growth.bag$importance[,1])

Importance.boost<-as.data.frame(summary(Growth.boost)[,2])

Importance$Average<-(Importance$Tree + Importance$Forest + Importance $Bagging + Importance$Boosting)/4

The corresponding code for the recession classification tree models is:

Importance.treeR<-100*as.data.frame(Recession.tree$variable.importance)/ sum (Recession.tree$variable.importance)

Importance.rfR<-100*as.data.frame(Recession.rf$importance[,4])/ sum(Recession.rf$importance[,4])

Importance.bagR<-100*as.data.frame(Recession.bag$importance[,4])/ sum (Recession.bag$importance[,4])

Importance.boostR<-as.data.frame(Recession.boost$importance)

ImportanceR$Average<-(ImportanceR$Tree + ImportanceR$Forest + ImportanceR$Bagging + ImportanceR$Boosting)/4.

We plot the partial dependence for the *randomForest* objects using the *partialPlot* function. There is no corresponding package in the *gbm* or *adabag* packages, so we have renamed the appropriate objects in the *partialPlot* function to create source code that will perform the same calculations to create partial plots for the Boosting and Bagging models[1]. We named our functions *partialPlotGBM partialPlotBoosting* for the *gbm* and *boosting* commands in the *gbm* and *adabag*

---

[1]We will make all these available on GitHub: https://github.com/bangecon/Growth-MachineLearning

packages, respectively.[2] We begin by loading these function into the work environment and subsetting the variable list to include the numerical variables in our dataset.

```
source("partialPlotGBM.R")
source('partialPlotBagging.R')
source('partialPlotBoosting.R')
Categoricals<-c("LagEthnicConflict", "LagNonethnicConflict", "System", "Execu-
    tive", "Regime")
Numericals<-rownames(Importance)[!rownames(Importance) %in% Categoricals].
```

To do the plots, we sequence through (seq_along, in fact) the numerical variables and plot the PDPs for the Bagging, Boosting, and forest models. We also run each model twice: the first run helps us to extract the maximum and minimum values on the y-axis for each algorithm's partial plot of the current variable in the loop; the second run plots all three algorithms (Bagging, Boosting, and forest) on one graph (with the y limits set as the highest maximum and the lowest minimum). The code also includes lines to save each finished plot as a jpeg and to add a legend to the graph. The loop for the growth regression tree models is:

```
for (i in seq_along(Numericals)) {
pdp.bag<-partialPlot(Growth.bag, pred.data = GrowthData.LS, Numericals[i],
    xlab=Numericals[i], main = paste("Partial Dependence on", Numericals[i]),
    plot = T, col = "blue")
lim.bag<-par('usr')
pdp.boost<-partialPlotGBM(Growth.boost, pred.data = GrowthData.LS, Numer-
    icals[i], xlab=Numericals[i], main = paste("Partial Dependence on", Numer-
    icals[i]), plot = T, col = "green", n.trees = 500)
lim.boost<-par('usr')
pdp.rf<-partialPlot(Growth.rf, pred.data = GrowthData.LS, Numericals[i],
    xlab=Numericals[i], main = paste("Partial Dependence on", Numericals[i]),
    plot = T, col = "red")
lim.rf<-par('usr')
yLimMin<-min(lim.bag[3], lim.boost[3], lim.rf[3])
yLimMax<-max(lim.bag[4], lim.boost[4], lim.rf[4])
jpeg(filename = paste(Numericals[i], ".jpg"))
partialPlot(Growth.bag, pred.data = GrowthData.LS, Numericals[i],
    xlab=Numericals[i], main = paste("Partial Dependence on", Numericals[i]),
    add = F, col = "blue", ylim = c(yLimMin, yLimMax))
partialPlotGBM(Growth.boost, pred.data = GrowthData.LS, Numericals[i],
    xlab=Numericals[i], main = paste("Partial Dependence on", Numericals[i]),
    add = T, col = "green", n.trees = 500)
```

---

[2]We have also created *partialPlotBagging* function for the *Bagging* function in the *adabag* package. It can also be found in the repository, but we do not use it for this book since we use *randomForest* to train the Bagging models here.

partialPlot(Growth.rf, pred.data = GrowthData.LS, Numericals[i], xlab=Numericals
    [i], main = paste("Partial Dependence on", Numericals[i]), add = T, col = "red")
dev.off()
legend("bottomright", c("Forest", "Boosting", "Bagging"), lty = c(1,1), lwd =
    1, col = c("red","green","blue"))}.

The analogous loop for the recession classification tree ensembles is:

for (i in seq_along(Numericals[1:25])) {
pdp.bagR<-partialPlot(Recession.bag, pred.data = GrowthData.LS, Numericals
    [i], which.class = "Recession", xlab=Numericals[i], main = paste("Partial
    Dependence on", Numericals[i]), plot = T, col = "green")
lim.bagR<-par('usr')
pdp.rfR<-partialPlot(Recession.rf, pred.data = GrowthData.LS, Numericals[i],
    which.class = "Recession", xlab=Numericals[i], main = paste("Partial Depen-
    dence on", Numericals[i]), plot = T, col = "red")
lim.rfR<-par('usr')
pdp.boostR<-partialPlotBoosting(Recession.boost, pred.data = GrowthData.LS,
    Numericals[i], which.class = "Recession", xlab=Numericals[i], main = paste
    ("Partial Dependence on", Numericals[i]), plot = T, col = "blue")
lim.boostR<-par('usr')
yLimMin<-min(lim.boostR[3], lim.bagR[3], lim.rfR[3])
yLimMax<-max(lim.boostR[4], lim.bagR[4], lim.rfR[4])
partialPlot(Recession.bag, pred.data = GrowthData.LS, Numericals[i], which.
    class = "Recession", xlab=Numericals[i], main = paste("Partial Dependence
    on", Numericals[i]), plot = T, col = "green", ylim = c(yLimMin, yLimMax), add
    = F)
partialPlot(Recession.rf, pred.data = GrowthData.LS, Numericals[i], which.class
    = "Recession", xlab=Numericals[i], main = paste("Partial Dependence on",
    Numericals[i]), plot = T, col = "red", add = T)
partialPlotBoosting(Recession.boost, pred.data = GrowthData.LS, Numericals[i],
    which.class = "Recession", xlab=Numericals[i], main = paste("Partial Depen-
    dence on", Numericals[i]), plot = T, col = "blue", add = T)
legend("bottomright", c("Forest", "Bagging", "Boosting"), lty = c(1,1), lwd =
    1, col = c("red","green","blue"))}.

One final note is that for each classification model, the *roc* function stores an "optimal threshold" that maximizes the sum of the sensitivity and the specificity. We can create variables for the predicted classes using these optimal thresholds by extracting the first element of *coords(roc object, "best")*.[3]

---

[3]The second and third element give the sensitivity and specificity at that threshold. Also, there are multiple objective functions on which we can evaluate the "optimal" threshold. One is the sum of the sensitivity and specificity (distance from the 45° line on the ROC curve); another is the closeness to the upper-left corner of the ROC curve.

 # Optimal Thresholds
Predicted Classes
GrowthData.TS$treeClassP <- ifelse(GrowthData.TS$treeProb < coords(roc.tree.
  GrowthData.LS, "best")[1], "NoRecession", "Recession")
GrowthData.TS$rfClassP <- ifelse(GrowthData.TS$rfProb < coords(roc.rf.
  GrowthData.LS, "best")[1], "NoRecession", "Recession")
GrowthData.TS$bagClassP <- ifelse(GrowthData.TS$bagProb < coords(roc.bag.
  GrowthData.LS, "best")[1], "NoRecession", "Recession")
GrowthData.TS$boostClassP <- ifelse(GrowthData.TS$boostProb < coords(roc.
  boost.GrowthData.LS, "best")[1], "NoRecession", "Recession")
GrowthData.TS$lmClassP <- ifelse(GrowthData.TS$lmProb < coords(roc.lm.
  GrowthData.LS, "best")[1], "NoRecession", "Recession")
GrowthData.TS$nnetClassP <- ifelse(GrowthData.TS$nnetProb < coords(roc.
  nnet.GrowthData.LS, "best")[1], "NoRecession", "Recession")
GrowthData.TS$averageClassP <- ifelse(GrowthData.TS$averageProb < coords
  (roc.ave.GrowthData.LS, "best")[1], "NoRecession", "Recession").

We can use the predicted classes from the optimal thresholds to reconstruct the confusion matrices for the models and re-estimate all of the measures of predictive accuracy. Using the optimal threshold will improve the measures of predictive accuracy *on average*, but it will often mean some measures will deteriorate.

We have submitted all of the code (and supplementary functions for the partial plots) to a GitHub repository that accompanies the book. These files are located at https://github.com/bangecon/Growth-MachineLearning

References

Sala-i-Martin, X. (1997). I just ran four million regressions. *American Economic Review, 87*, 178–183.
Temple, J. (1999, March). The new growth evidence. *Journal of Economic Literature, 37*, 112–156.

© The Author(s) 2017
A. Basuchoudhary et al., *Machine-learning Techniques in Economics*,
SpringerBriefs in Economics, https://doi.org/10.1007/978-3-319-69014-8